ECONOMIC COMMISSION FOR EUROPE
Geneva

Air Pollution Studies 4

EFFECTS AND CONTROL OF TRANSBOUNDARY AIR POLLUTION

*Report prepared within the framework of
the Convention on
Long-range Transboundary Air Pollution*

UNITED NATIONS
New York, 1987

NOTE

Symbols of United Nations documents are composed of capital letters combined with figures. Mention of such a symbol indicates a reference to a United Nations document.

*

* *

The designations employed and the presentation of the material in this publication do not imply the expression of any opinion whatsoever on the part of the Secretariat of the United Nations concerning the legal status of any country, territory, city or area, or of its authorities, or concerning the delimitation of its frontiers or boundaries.

ECE/EB.AIR/13

UNITED NATIONS PUBLICATION

Sales No. E.87.II.E.36

ISBN 92-1-116410-9

02000P

Table of Contents

INTRODUCTORY NOTE AND SUMMARY

This fourth volume of the series of Air Pollution Studies, published under the auspices of the Executive Body for the Convention on Long-range Transboundary Air Pollution, contains seven technical reports on effects of air pollution and on technologies to control emissions.

The general background to the Convention and its progress up to 1986 was described in the introduction to volume 3 of this series, *Transboundary Air Pollution : Effects and Control*, (Sales No. E.86.II.E.23) and in *National Strategies and Policies for Air Pollution Abatement* (Sales No. E.87.II.E.29).

During 1986, work on implementation of the Convention and its related protocols continued under the auspices of the Executive Body and subsidiary bodies, including the newly established Working Group on Nitrogen Oxides. In addition to the Co-operative Programme for Monitoring and Evaluation of the Long-range Transmission of Air Pollutants in Europe (EMEP), with 90 monitoring stations in 24 countries, there are now three international co-operative programmes in operation. They monitor and assess the effects of air pollution on forests, on rivers and lakes, and on materials, including historic and cultural monuments. Scientific and technical information was exchanged at a number of expert meetings and workshops organized or hosted by Parties to the Convention, including the International Conference on Acidification and its Policy Implications (organized in Amsterdam in May 1986 by the Government of the Netherlands in co-operation with ECE) and the Fourth Seminar on the Control of Sulphur and Nitrogen Oxides from Stationary Sources (hosted by the Government of Austria in Graz in May 1986).

The present technical reports - several relate to or update earlier studies published in this series - are the result of the continuing review of scientific and technical knowledge undertaken in the context of the Convention. Apart from necessary editing, care has been taken to avoid any substantive change in the reports as reviewed by the Executive Body for the Convention at its fourth session in November 1986.* Sole responsibility for the text rests with the secretariat of the Economic Commission for Europe.

Effects of Air Pollutants on Vegetation and Soil

While previous reports dealing with effects on vegetation focused on crops, this study concerns mainly air pollution effects on forests. In 1984, the area of forest damaged in 10 countries of Europe where surveys were conducted was estimated to be equivalent to about 6 million hectares. This represents about 24 per cent of the total forested area of these countries.

Cause-effect relations and dose-response functions relating to air pollution impacts upon vegetation and soil are complex and varied. The main variables are environment, response of individual organisms, sensitivity of populations, communities and ecosystems as well as other biotic, edaphic or climatic stresses.

Within a conceptual framework for the evaluation of cause-effect and dose-response relations, the study investigates direct effects on vegetation as cellular, biochemical and physiological effects and as effects on whole plants and crops. Critical levels are discussed for single pollutants and for combined effects of various gaseous pollutants, and direct interactions with biotic stresses. Cause-effect relationships and dose-response functions are also associated with: (a) changes in soils induced by air pollution; and (b) combined direct and indirect effects in complex systems, notably forests.

---

* Chapter 1: EB.AIR/WG.1/R.4 and Corr.1; chapter 2: EB.AIR/WG.1/R.15 and Corr.1; chapter 3: EB.AIR/WG.1/R.17 and Corr.1; chapter 4: EB.AIR/WG.1/R.25; chapter 5: ENV/WP.1/R.69/Rev.2 and Corr.1; chapter 6: ENV/WP.1/R.77/Rev.1 and Corr.1; chapter 7: ENV/WP.1/R.86 and Corr.1.

Effects of Air Pollutants on Aquatic Ecosystems

Freshwater acidity is a result of complex interactions between wet and dry deposition, soil and rock conditions, land use and the hydrology of each catchment. Hydrological conditions and biogeochemical processes affect run-off composition and consequently the quality of freshwater. The study summarizes new developments, with a view to clarifying the role of air pollution in the acidification of surface waters.

A relationship between acid deposition and soil acidification has been established in many areas. There is also evidence that this acidification is directly related to negative effects in a number of areas of Europe and North America.

Most freshwater plants and animals have a specific pH-tolerance range. They disappear from a system when pH exceeds this range. In addition to such direct effects, organisms may disappear following the loss of their plant food or prey species.

Models have recently been developed both to help understand acidification processes and to predict the consequences of changed emissions. Several models, briefly summarized in the study, have important common features with respect to soil chemical processes.

Mechanisms and Effects of Water and Soil Acidification on Structures

Corrosion of materials in contact with soil and water is a complex problem. The corrosion rate and type of corrosion are determined by numerous factors. So far only the corrosion of structures above ground, mainly by sulphur pollutants, has been taken into account. In view of the ongoing acidification of surface waters and soil and the risk of acidification of ground water, the influence of corrosion on buried installations and installations in water also calls for attention.

Effects of soil acidification are studied in relation to carbon steel, cast iron, zinc, galvanized steel, copper, lead, concrete and impregnated wood. Many essential structures are buried in soil. A number of technically and economically important structures are potentially susceptible to soil acidification: water mains of steel and cast iron; conduits of galvanized steel, copper and lead connecting mains to houses; road culverts of galvanized steel and concrete; lead-sheathed telephone cables; oil and gasoline storage tanks; powerline tower foundations including stays; and concrete sewers.

Internal corrosion of water pipes, whether copper, lead or galvanized steel, has become a great problem today in areas sensitive to acidification, especially where there is no corrosion control of water supplies. Where lead pipes or solder containing lead are used, the elevated lead content in tap water also creates a potential health hazard. Increased corrosion of road culverts can also be expected in acidified areas.

There are major gaps of knowledge in this field. The need for research comprises, among other aspects, assessment of the effect of acid deposition on soil and water pH in different regions, studies of the mechanisms of corrosion from acidification by sulphur and nitrogen compounds, establishment of dose-response relations and cost estimates for damage from acidification.

Damage to Materials by Air Pollution: Methods for Assessing Stock-at-risk

The determination of the amount of exposed material susceptible to acid deposition is known as the materials inventory. Basically two approaches may be taken in preparing materials' inventories. One is a true inventory, or census, which enumerates each building/structure and provides an individual record that describes its major features. The damage is calculated for each building and then added up to provide the total damage for all buildings in the inventory.

The other approach proceeds from first totalling all the areas of exposed materials and then computing the damage by using an average damage-function factor applied to this total. In this second approach, the identity of individual buildings is lost, and the materials inventory is characterized by a probability function for the exposed material per unit area of land. The second approach is sometimes referred to as a materials distribution assessment, so as to emphasize that it is not a true building-by-building inventory.

The choice between these two approaches depends upon the structure of the intended cost-benefit analysis and usually reflects the dichotomy between cultural and utilitarian architecture.

Specialized inventories are also possible. The locations of bridges and electrical transmission towers is a matter of record. Building-by-building records are often available for central business districts. In such cases, separate inventories may be more useful than trying to predict materials' distribution from a general model.

In order to put into perspective the physical damages computed from the deposition rates and the damage functions, it is necessary to convert them into changes in service life. This is a way of providing a common yardstick for comparing damages to different materials, or to the same material used in different applications.

By developing damage functions, service lives and materials inventories, an estimate can be made of the physical damage caused by air pollution. However, the ultimate goal is to place an economic value on these damages in order to weigh them against the costs of control.

Control of Sulphur Dioxide Emissions from Industrial Processes

Although the largest contributors to anthropogenic SO_2 emissions are utility and industrial boilers, significant amounts of SO_2 can also originate in industrial processes such as non-ferrous metal smelting, petroleum refining, cement manufacturing, iron and steel manufacturing, and pulp and paper production.

Efforts to control air pollutant emissions from non-ferrous metal production processes have traditionally focused on the problem of SO_2 control. Copper-, nickel-, lead- and zinc-processing and control technologies are presented in the study.

The petroleum-refining industry employs a wide variety of processes to convert oil into more than 2,500 refined products; these processes use catalytic cracking (fluidized-bed catalytic cracking, moving-bed catalytic cracking), thermal cracking (visbreaking, coking), utilities plant (supplying the steam necessary for the refinery), process heaters and sulphur recovery plants.

Emissions of sulphur oxides from cement kilns are predominantly SO_2. The sulphur content of the fuel and raw material feed both contribute to the total SO_2 generated in the kiln. However, only a fraction of the SO_2 generated in the kiln is actually released to the atmosphere (10 to 25 per cent). If further SO_2 control is needed, scrubbers can potentially achieve a 90 per cent reduction in SO_2 emissions.

For iron and steel manufacturing, different processes and related control technologies are presented: Sintering, coking and production of ferro-alloys.

The pulp and paper industry is included, with sulphate and sulphite processes and related control technologies.

Emission Control by Fluidized Bed Combustion (FBC)

In recent years the use of high quality fuels for power and heat generation has diminished significantly owing to rapid increases in their prices. Conversely, the availability of alternative low-grade fuels, such as lignite, coal, and various kinds of industrial and municipal waste, has led to their wider use.

Combustion of these fuels in conventional boilers can present serious environmental problems. These include flue gas emissions of SO_x, NO_x, particulates, and trace elements, some of which are highly toxic. Considerable research and developmental work has been carried out worldwide in the field of fluidized bed combustion as a means of overcoming the environmental problems associated with high-sulphur fuels in particular.

The following types of fluidized bed combustion plants were included in the study: bubbling bed, circulating bed (CFBC), multi-bed, pressurized fluidized bed. Fluidized bed technology has been successfully developed into a commercially viable alternative to other combustion technologies. Developmental work is continuing on the different bed types in the search for even more cost-effective and environmentally sound energy production plants that can use low-grade fuels. Although the largest plants presently in operation can be classified as medium-sized, there is no foreseeable reason why FBC technology should not be incorporated into large projects in the future. Emission control in FBC plants has proven to be less difficult and less expensive than in conventional plants.

Inventory of Technologies for Controlling Emissions of Nitrogen Oxides from Stationary Sources

The conclusions of this task force report review the current state of technologies for reducing nitrogen oxide emissions from stationary sources, with special emphasis on practical experience, and costs of these technologies. In view of the fact that NO_x emissions from stationary sources arise primarily during combustion of fuel for generation of heat and electricity, the focus is on a description of NO_x reduction technologies in large-scale power plants (i.e. power plants with a rated capacity of more than 50 MW_{th}).

Combustion modifications (primary measures) for avoiding the formation of nitric oxide are described as well as methods for flue gas treatment (secondary measures) to reduce nitric oxide emissions. Besides combustion modifications and flue-gas treatment technologies - which have already been tested on a large scale - the conclusions also refer to measures which may become available before 1990. Control measures for other NO_x emission sources are outlined as well. Although such sources may be of secondary importance in many countries in terms of total national emissions, they can nevertheless contribute substantially to local/regional air pollution problems.

In addition to the technical description and assessment of different NO_x control technologies, the conclusions outline the investment requirements and operating costs of each measure as far as available data permit. The state-of-the-art of different control measures in various countries is examined, and trends are discussed.

Part ONE

Effects of Air Pollutants

Chapter 1

EFFECTS OF AIR POLLUTANTS ON VEGETATION AND SOIL: CAUSE-EFFECT RELATIONSHIPS, WITH SPECIAL REFERENCE TO AGRICULTURE AND FORESTRY

1. Adverse effects of air pollution have long been recognized (8, 10). Queen Elizabeth I of England promulgated laws in the seventeenth century in order to control obnoxious fumes from coal fires in London. Rudimentary controls of industrial emissions were eventually introduced there and in a number of other industrialized countries by the nineteenth century, the resulting benefits to vegetation being noted. By then, scientists had begun to investigate and express concern about possible serious damage to agricultural, horticultural, forest and ornamental plants.

2. Investigations up to the 1950s focused on the appearance of leaf necrosis, shoot die-back and plant death - acute visible damage - attributable to large concentrations of air pollutants which rarely occur nowadays. The results of early investigations helped to establish some of the principles of cause-effect relationships which are still relevant to current air pollution problems.

3. With the drop in concentrations of pollutants at ground level near sources of emissions following the introduction of clean air acts, attention then switched to damage at the biochemical and cellular level in the absence of obvious blemishes - in other words, chronic damage. This stimulated interest in the complex interactions between air pollution, climate, vegetation, soils and freshwater. Public and scientific concern has subsequently grown apace as evidence of possible major environmental changes attributable wholly or in part to air pollution has emerged from more and more countries in the northern hemisphere (5, 10). Much of this attention is focused on the need to identify and quantify cause-effect relationships for the development of rational and effective measures to control air pollution and protect the environment (11, 12).

4. The case for controlling emissions of atmospheric pollutants is based essentially on proof of a relationship between cause (presence of a pollutant or pollutants) and effect (e.g. damage to vegetation) (9). This case is enhanced if the cause-effect relationship can be quantified in terms of a dose (amount of pollution per unit of time)-response (level or intensity of effect) function. Dose-response functions also have great value in the prediction of air pollution impacts and in their economic assessment (9, 12 to 15). (The word "dose" is used here in the sense of potential dose, i.e. concentration multiplied by duration of exposure; and not as effective dose which is the amount of the pollutant actually absorbed.)

5. The scientific literature on cause-effect relations and dose-response functions relating to air pollution impacts upon vegetation, soils and freshwater is voluminous. However, recent reviews (see references 1 to 32) have revealed serious deficiencies in information, partly owing to lack of research and partly to the inherent complexity and variability of cause-effect relationships. The primary variables are:

(a) **The pollution environment** - The atmosphere contains a mixture of natural and pollutant components which vary greatly in time and in space.

(b) **Individual organisms** - Each pollutant induces a different but often overlapping set of responses within an organism. These responses may all be related or only partly. Thus they will have different consequences for the development, growth and survival of the organism, and for its habitat.

(c) **Populations, communities and ecosystems** - There is considerable genotypic variation in the sensitivities and responses of individuals within a population of organisms. This source of variation is enhanced within communities by the inherent differences in sensitivity between species. Cause-effect relationships become exceedingly complex at the ecosystem level where a single pollutant may travel a number of different pathways and induce both direct and indirect responses in the constituent organisms. Moreover, tolerant species may take competitive advantage of the niches vacated by more sensitive members of the community.

(d) **Other stresses** - The cause-effect relationships between a pollutant and damage can be modified greatly by other stresses which themselves may produce symptoms like those caused by pollutants. These stresses may be biotic (e.g. disease), edaphic (e.g. soil type,

mineral nutrition) or climatic (e.g. rainfall, temperature, wind exposure). They can alter exposure, and intensify responses (e.g. reduced growth). Drought-induced moisture stress and frost are perhaps the best understood of the "modifiers" (33).

6. Further complications arise from the fact that the sources of variation in cause-effect relationships are not entirely independent of each other. Thus, high winds increase exposure to gaseous pollution and increase moisture stress which in turn may partially close stomata, thereby reducing pollutant uptake but also amplifying moisture stress by inhibiting root development in plants (8, 16). Such issues are discussed in greater detail below in relation to forestry and agriculture.

I. THE POLLUTION ENVIRONMENT

7. This section summarizes current information on trends and concentrations of ambient air pollutants relevant to cause-effect studies. It must be emphasized that mean concentrations over large regions have only limited relevance. The researcher is generally much more interested in concentration and deposition values (a) in and around urban/industrial areas, and (b) in rural areas at increasing distances from significant conurbations/industrial sources. Determination of dose-response functions requires even more detailed and site-specific information.

A. EMISSIONS

8. Comparisons were made of samples from the Antarctic and Greenland ice sheets. They indicate that the current anthropogenic input of sulphur and total acid compounds to recent snow in Greenland is between one and two times greater than the assumed natural input in Antarctica. The origins of the natural input are uncertain but probably include volcanic emissions and fluxes of sulphur gases from the oceans. The enhancement of inputs to Greenland implies wholesale contamination of the northern hemisphere; comparable data exist for lead and other metals (10, 19, 34).

9. Trends in emissions of primary air pollutants in the northern hemisphere exhibit great variation, but certain general conclusions may be made (3, 8, 10, 19). Emissions of sulphur dioxide (SO₂) increased between 1950 and 1970

by a factor of 3 in most of Europe and parts of North America, but then stabilized. In the eastern United States of America, however, the increase was much greater and continued into the early 1980s while, in contrast, emissions markedly declined in some parts of western Europe. Between 1950 and 1980 there was also a general shift from low-level domestic to high-level stack (energy and other industries) emissions, reducing the local concentration of gaseous pollutants. During the same period of time there have been marked reductions in particulate emissions including those of bases.

10. The data on nitrogen oxides (NOₓ) are less comprehensive, but the trends probably followed those of SO₂ up to 1970. From then on, emissions appear generally to have increased with the growth of internal combustion engines, into the 1980s. The increases in hydrocarbon emissions have been larger. Again, however, there are regional variations. For instance, increases were greatest in the Federal Republic of Germany in the 1980s but stabilized in the United Kingdom during the same period. However, they are expected to increase further over the next decade in Europe (35). Emissions of lead and other metals may have followed the same pattern up to the mid-1970s, but then stabilized or declined as control measures were introduced in some countries.

11. The changes in the quantities and types of emissions since the start of the industrial revolution have greatly influenced the transport and deposition of pollutants in the northern hemisphere. The emphasis has shifted from local

problems of acute particulate and SO_2 pollution, to long-distance and widespread acid deposition; the latter being reinforced by the growth of NO_x emissions. NO_x and hydrocarbon emissions have contributed to an increasing incidence of widespread secondary ozone (O_3) and other photo-oxidant pollutants, hydrocarbon concentrations often being the rate-limiting factor in their production. Increasing amounts of ammonia (NH_3) pollution are also reported. Loadings of larger particles, on the other hand, have noticeably decreased in most areas.

B. GASEOUS POLLUTANTS: CONCENTRATIONS AND DRY DEPOSITION

12. Urban concentrations of SO_2 typically range between 20 and 60 ppbV (50-160 ug m^{-3} air) daily annual mean values in the United States of America and in Europe. In the United Kingdom, such values have declined by 40 per cent over the past decade (19). The dry deposition values including fine particulate aerosols of 0.1 to 1.0 um diameter, can be as high as 200 kg SO_2 ha^{-1}yr^{-1} in industrial heartlands. Values for NO_x, including NO_2 and NO, are not dissimilar in western Europe at 20 to 50 ppbV, corresponding to a dry deposition of 10 to 25 kg N.ha^{-1}yr^{-1} (including the deposition of vaporous nitric acid), but may reach 100 to 300 ppbV in central Europe and North America. Annual mean daily values for O_3 are largely misleading. In industrial regions, background concentrations of ozone are 20 to 40 ppbV (40 to 80 ug.m^{-3}), sometimes increased to episodes of a number of hours with 150 ppbV or more. The number of days above 60 ppbV, and the peak values, are considerably greater in parts of North America (8, 18). Other gases such as NH_3 and hydrogen fluoride (HF) rarely rise above 10 ppbV. Particulates including ammonium sulphate aerosols range between 10 and 100μ m^{-3} air including the equivalent of 5 to 7μ SO_2 m^{-3} air.

13. In rural areas near cities, concentrations of SO_2 and NO_x range from 2 to 40 ppbV, NO_2 accounting for 70 to 80 per cent of the NO_x. Particulates range between 2 and 30μ m^{-3} air, while O_3 concentrations are similar to the urban values. In rural areas, concentrations are known with less certainty but the mean annual values for SO_2 are 0.1 to 10 ppbV while NO_x is between 0.2 and 10 ppbV. The number of "O_3 days" above 60 ppbV ranges from 5 to 20 per year with even larger numbers at high altitude than at low altitude. Particulate values scarcely differ from those recorded closer to urban areas.

C. TEMPORAL AND SPATIAL VARIATIONS IN DRY DEPOSITION

14. Atmospheric concentrations of SO_2 vary diurnally and seasonally. They are usually larger late in the morning than at other times of the day; amounts in winter usually exceed those during summer. The concentrations of NO_x follow similar patterns with winter peak concentrations occurring sooner, in November/December rather than January/February. The patterns for ozone are markedly different. The diurnal peak concentrations of ozone occur during the afternoon but the daily changes are markedly less at high, than at low altitudes: seasonal maxima occur in April-July with minima in December/January. The transient concentrations of phytotoxic gaseous peroxyacetyl nitrate (PAN) tend to parallel those of ozone with peak hourly concentrations of 20 ppbV being detected in Europe (18). Similar, but not necessarily identical, patterns of pollutant occurrence have been found in the Federal Republic of Germany, the Netherlands, the United Kingdom and in North America.

15. The peak short-term (daily and/or hourly) values for the above pollutants over afforested and agricultural areas are of special interest. Maximum hourly values of 40 ppbV for SO_2, NO_x or NO_2 are common as are O_3 peak values of 100 to 400 ppbV. It is, of course, risky to extrapolate from the limited data available at a few sites to infer values for large areas of a country. However, estimates from the United Kingdom suggest that 10 million hectares are subject to mixtures of SO_2 and NO_2 with an annual mean of 10 and 15 ppbV, respectively, with the attendant diurnal and seasonal fluctuations. About 2.3 million hectares are exposed to 20 ppbV SO_2 and 20 ppbV NO_2 and 0.23 million hectares to 40 ppbV of each gas (19). The incidence and spatial distribution of two or three consecutive "O_3 days" are much more variable, depending on weather conditions. Again, similar but not identical extrapolations have been attempted in several countries including the Federal Republic of Germany, the Netherlands and the United States of America (primarily for O_3) (13, 15, 36). Attempts have been made to use national and EMEP data for similar purposes in western Europe (36, 37, 38). The approach in the United States of America was interesting because it used a surrogate data set for rural areas to compensate for the bias towards urban monitoring in the Environment Protection Agency's SAROAD data system, and to generate average 7-hour daily ambient O_3 concentrations for each year between 1978 and

1982, on a regional basis (i.e. the concentration of O_3 during the active period of synthesis) (13). The days when O_3 exceeded or equalled 60 ppbV were not calculated, but the "background" level of 40 ppbV was clearly exceeded in most regions between April and September in two or more years of the period 1978 to 1982.

D. WET DEPOSITION

16.   Unlike dry deposition, wet deposition, excepting that of ammonia, is not closely related to local sources. Natural sources of sulphur-, nitrogen-, chlorine-containing gases and other acids in the atmosphere produce annual average acidities in rain and snow of ≥ pH 5.0. This degree of acidity is not greatly exceeded on the north-western periphery of Europe, but it is exceeded by a factor of 7 (around pH 4.2) over many areas of continental Europe and north-eastern North America. The above pH values equate respectively with approximately 10 u eq and 70μ eq H^+ l^{-1}.

17.   These mean acid values conceal a systematic seasonal cycle when rain in Canada, the Scandinavian countries and the northern part of the United Kingdom is more acid in April/June than at other times of the year. There is also a significant spread in the amounts of acidity in different rain events, a mean pH of 4.2 hiding extremes of pH 2.9 to 7.0 (41). Nitric acid and sulphuric acid contribute 30 to 40 per cent and 60 to 70 per cent, respectively, to the total acid in wet deposition. The relative contribution of nitric acid has increased over the last 30 years, in some cases doubling. The acidifying potential of wet deposits is also influenced by the concentrations of ammonia and base cations. The rate of wet deposition of sulphur ranges from less than 3 kg S $ha^{-1}yr^{-1}$ in northern Scandinavian countries to more than 30 kg S $ha^{-1}yr^{-1}$ in rural areas of central Europe. The corresponding figures for inorganic-nitrogen are in the range of 1-3 kg N $ha^{-1}yr^{-1}$ in northern Scandinavian countries compared with measured deposition values of more than 20 kg N $ha^{-1}yr^{-1}$ over large parts of central Europe and North America. Ammonia can also greatly influence the acidifying potential of rain;   although it neutralizes acids in rainwater, ammonia is eventually converted into nitrate by nitrifying bacteria in the soil. This process produces two H^+ ions for each NH_4^+ ion nitrified.

18.   There is also wet precipitation in the form of fog, mist and low cloud - "occult" precipitation.   While more common at higher than lower altitudes, its contribution to wet deposition and total deposition is difficult to measure accurately.   Nevertheless, analyses of samples taken in many areas of Europe and North America indicate that "occult" precipitation contains much higher concentrations of most ionic species including H^+ than does ordinary rain (10). There are also indications that the condensation nuclei of many forms of "occult" precipitation originate from anthropogenic emissions of NH_4^+, fly ash and other fine particulates (42).

E. TOTAL DEPOSITION

19.   It will be clear from the preceding discussion that the relative contributions of wet and dry deposition to total deposition of a particular pollutant or ionic species also vary in time and space. Thus, the ratio of dry to wet deposition of sulphur ranges typically from about 4:1 in industrial regions with major emitters of SO_2, to 1:4 in areas towards the periphery of long-range transport (e.g. Scandinavian countries). Thus, patterns for total depositions of sulphur, nitrogen and even hydrogen over agricultural and forestry regions are highly heterogeneous. As a rough guide, the deposition of sulphur over central Europe is in the range of 30 to 60 kg S $ha^{-1}yr^{-1}$; values for nitrogen would be of the same order (8).   Concentrations are larger in winter and spring than in summer. They decline towards the western and northern coastal areas.

20.   Deposition of heavy metals has not been studied to the same extent. However, deposits of four metals (Cd, Mn, Zn, Pb) in rural sites of the Federal Republic of Germany were significantly higher than those for equivalent sites in the eastern United States, which in turn were higher than those in Greenland (10, 34).   Other data suggest that there is a gradient of increasing deposition of metals and elements of non-marine origin from western, towards eastern Europe (43). There may also be a south-west to north-east gradient of increased deposition of fine particulates acting as aerosol nuclei.

F. PATHWAYS

21. Pollutants deposited from the atmosphere may influence plant growth and development through various combinations of multiple pathways and mechanisms, and over widely varying time scales. Dry deposition of gases is greatly influenced by two sets of resistance: atmospheric and surface.   Of these, surface

resistance is usually appreciably stronger than atmospheric resistance, the differences being greatest for tall and aerodynamically "rough" vegetation such as forests, where atmospheric resistance is relatively slight. Surface resistance has four components: stomata, leaf surface (or cuticular), surface water, and soil (44). Of these, stomatal resistance is by far the most important because stomata are the main pathways of gases to and from the plant. However, some gases enter through the cuticle, with or without modification, when surface moisture is present.

22. These pathways may be used by some pollutants which are soluble in surface water, and by pollutants dissolved in wet precipitation. However, some gases and most particulates deposited dry remain on the leaf surface as part of the surface burden which is thus a complex mixture of pollutants deposited by different processes and undergoing various stages of transformation (e.g. SO_2 to SO_4^{2-}). In addition, the canopy may exchange material with the percolating rain which thereby is enriched in some elements and depleted in others. The exchange processes and the fate of particular ions in throughfall vary with species, age, site and nutritional status of the trees. In addition they

vary with the acidity of the rain and the amount of dry deposition. Most species have a net loss of K^+, Ca^{2+}, Mg^{2+} and SO_4^{2-}. The effect on H^+, NH_4^+ and NO_3^- varies, but pH often decreases in throughfall from conifer species and increases from broadleaved species. Stemflow is often more acidic than throughfall. The effect of foliar leaching on the nutritional status of the trees is not completely known as several studies indicate that nutrients leached are replaced by uptake from the soil.

23. Pollutants and leachates reaching the soil in throughfall and stemflow combine with those deposited directly upon the soil surface to form the total input to the soil system. These inputs may have a wide range of direct and indirect effects upon the soil itself, upon plant growth and development, and upon the composition and quality of soil drainage waters entering aquifers and surface waters. For instance, inputs of H^+ ions are exchanged for basic cations (mainly Ca) which, unless immobilized, are rapidly leached from the soil. Elevated inputs of nitrogen can contribute to both soil acidification and fertilization. These issues are discussed later (section IV.A).

II. CAUSE-EFFECT RELATIONSHIPS

24. The previous section briefly reviewed the ambient air pollution environment of vegetation in agricultural and forest systems. Later the evidence of cause-effect relationships between ambient air pollution and direct damage of indirect effects in terrestrial systems will be examined. First, it is necessary to establish a conceptual framework for the evaluation of cause-effect and dose-response relations.

A. THEORETICAL CONSIDERATIONS

Attribution of cause

25. Cause-effect relationships of air pollutants on terrestrial and aquatic ecosystems are known to be complex and strongly influenced by soil type, vegetation, climate, etc. Although absolute proof based on the application of Koch's postulates (8, 11 and 46) should be the aim, it is unlikely to be achieved for many years. In the interim, use must be made of the best available data.

26. Regrettably, the rigorous criteria propounded by Koch have frequently been

ignored when investigating the effects of air pollutants. He indicated the desirability of reproducing symptoms of disease or, as far as atmospheric pollutants are concerned, ecosystem impairment, in order to link effects to causes.

Dose-response functions

27. The basic principles of dose-response functions have been established over the past 50 years by toxicologists and epidemiologists working with very wide ranges of organisms - humans, animals, plants, microbes - and toxicants including atmospheric pollutants, albeit at concentrations now regarded as excessive compared with ambient levels of pollution (9, 47). Nevertheless, the relatively limited data for chronic and ambient exposures indicate that the same basic principles apply, namely:

(a) Different doses of a pollutant produce a range of responses in a target organism, the responses not necessarily being related in time and space;

(b) The relative order of sensitivity of a range of target organisms is often different for:

(i) two responses to the same pollutant; and (ii) the same response to two different pollutants;

(c) The period of exposure for appearance of damage increases as the concentration of pollutants decreases. In the example shown (Figure 1), an exposure of 60 minutes is required at an ozone concentration of 400 ug to produce damage whereas an exposure period in excess of 12 hours is required when the O_3 concentration is only 80 ug. The data in Figure 1 are for a wide range of plants and absolute values will inevitably differ from species to species. Figure 2 shows the relationships between the exposure period and concentration of PAN and confirms the generalizations already made for O_3.

(d) Figures 1 and 2 also demonstrate the concept of threshold values or critical limits. For example, Figure 1 shows that at O_3 concentrations of less than 60μ , symptoms will never appear irrespective of the exposure time. On this basis 70μ can be regarded as the threshold value or critical limit.

(e) The relationship between increasing dose and increasing magnitude of a response - the dose-response function - is usually non-linear; sigmoidal or quadratic functions being common; and

(f) Changes in parameters other than concentration and duration of exposure (e.g. nutrition, temperature) can alter the threshold limits and the dose-response function.

28. Principle (f) is especially troublesome and has been quantified mathematically in very few cases. Figure 4 is an example of a modifying function for moisture stress interacting with O_3 pollution on crop yield. Further complications ensue from exposures involving:

(i) Mixtures of pollutants whose combined action may be less than (antagonistic), equal to (additive), or greater than (synergistic) the sum of their independent effects (Figures 3 and 5).

(ii) Indirect effects (e.g. modifications of soil fertility) or combinations of direct and indirect effects (10).

(iii) Ecosystems composed of species of widely differing forms, life cycles and sensitivities; and

(iv) Ambient levels of pollution often being close to the threshold limits or in the lower ranges of the dose-response function where the magnitude of the response is minimal and highly sensitive to modifying influences (Figure 4).

29. It is easy to understand from these points, therefore, that the mathematical expression of the dose-response function is extremely complex and very specific to the individual target organism and/or system.

B. THE ECOLOGICAL CONTEXT

30. The environment is not a stable entity but a dynamic system in time and space. There is constant change and evolution as nature tries to establish the optimum balance between all environmental forces and stresses, including the activities of Man. It will be clear, however, that for ecosystems containing long-lived species such as trees, adaptation cannot keep pace with relatively fast changes in stress factors, such as air pollution.

Ecological succession

31. The theory of ecological succession states in essence that an ecosystem will evolve through a series of successive stages to the optimum state consistent with prevailing biological, climatological and edaphic conditions. Thus, lowland pastures on temperate brown earths would be expected to evolve through scrubland to mixed deciduous forest dominated perhaps by oak, if left to their own devices. In the optimum state or climax, "mature" or "over-mature" specimens of the dominant species would die, thus releasing nutrients and making way for their niches to be occupied by competing younger individuals of the same or another co-dominant species.

32. Evidence in support of ecological succession and maintenance of the climax states can be found in the palaeobotanical record, and in the examples of self-sustaining and self-regeneration systems of apparently great age (e.g. some old forests). It is also significant that the theory predicts the death of "over-mature" specimens and allows for the impact of climatic change (9, 10, 50, 51).

Figure 1

The exposure (daily ozone concentration multiplied by time) at which metabolic activity, enzymatic and macroscopic growth are significantly decreased in a range of plant species.

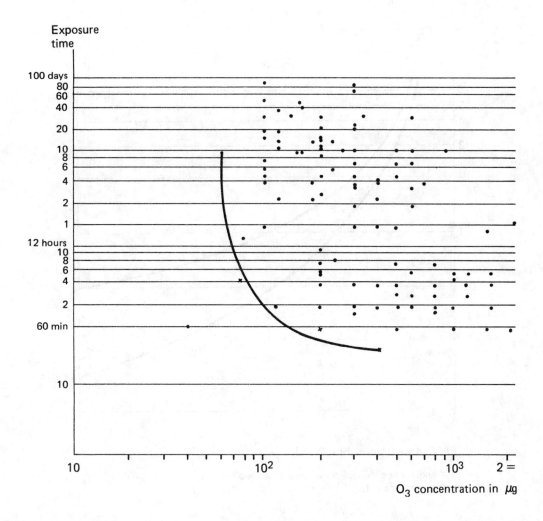

Figure 2

Two lines of lowest effective exposure to PAN for plants (18)

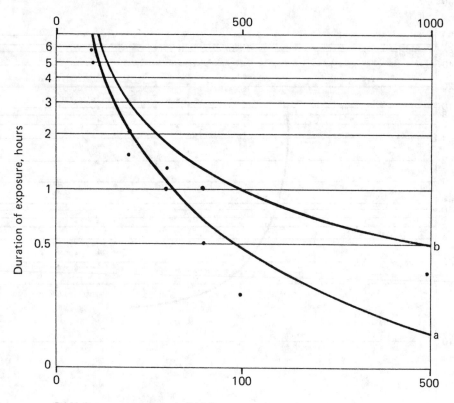

PAN Concentration, ppb (V/V)

a = aggregate of selection of sensitive species
b = Urtica urens L

Figure 3

Percentage yield reduction in Dutch agricultural production related to air pollution levels (15)

The 1983 levels are used as a reference (100).taking natural background levels as 0%

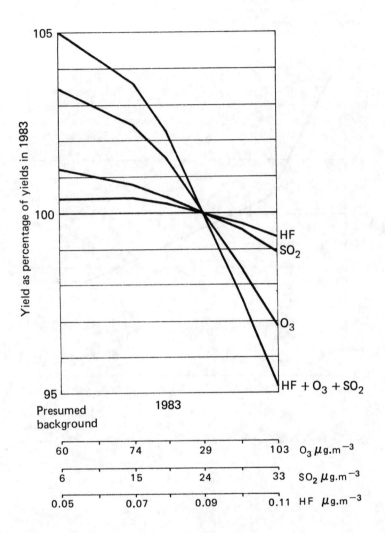

Figure 4

Decrease in response of crop (soya bean) when exposed to ozone at different moisture stresses

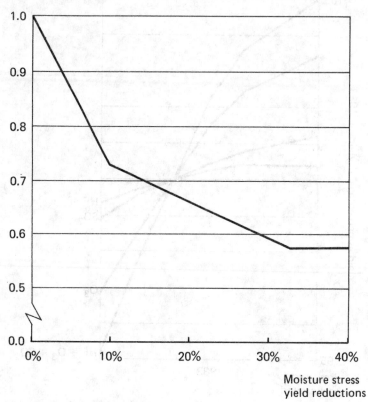

Source: King (1984) (33)

Figure 5

Variation in response at the lower ranges of dose ppbV SO_2^{-1} 24 hr^{-1}

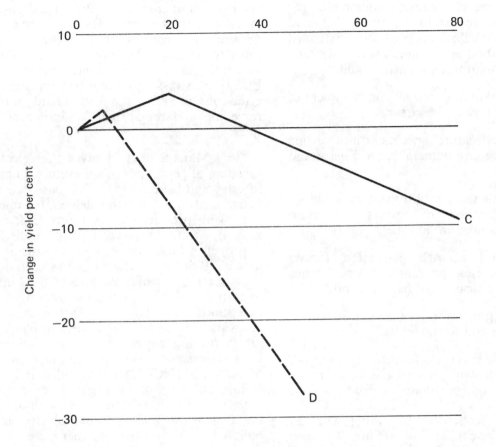

Line C = Function derived by Roberts for ryegrass, SO_2 only
Line D = Function derived for rye grass from Mansfied & Freer–Smith, Whitmore, etc., for
plants grown in ambient air polluted by SO_2 and NO_2 indicating possible synergism.

Climatic change

33. Evidence of climatic change also comes from the palaeobotanical record, and in recent times, from meteorological and dendrological studies (10, 52). Baumgartner (52) summarized the possible effects of seemingly minor changes in climate upon temperate forests. Minimal changes in mean values coupled with extended extreme events such as droughts, severe winters, early frosts and mild/wet summers would:

    (a) Initiate the retreat or advance of individual species over wide areas;

    (b) Force "mature" specimens into decline and make them more vulnerable to disease and deficiencies;

    (c) Inhibit the establishment of seedlings and the regeneration of forests if certain minimum conditions were not met;

    (d) Place at risk plantation forests composed of species, provenances and clones planted outside their original habitats; and

    (e) Have greatest impact upon tree species at their montane and Arctic limits.

34. The references to "mature specimens", "decline" and "areas of greatest risk" bear many similarities to the postulated effects of air pollution on forests (10). These issues will be taken up later (section III.B), along with an examination of possible interactions between pollution and climatic changes as the cause of forest damage and decline. Other natural ecosystems such as tundra, mountain grasslands and wetlands may be affected by climatic change over long periods of time.

Management

35. Most terrestrial and freshwater ecosystems of the northern hemisphere are managed to some degree. Regimes range from light sanitation activities in order to maintain the health and vigour of the system, to the intensive regimes of lowland monocrop cultivation. However, the objectives of management remain the same, namely to arrest ecological succession in a desired state and to modify and/or optimize the fertility and productivity of the system for a particular purpose. Management, therefore, has profound influences upon the responses of organisms and ecosystems to stress (5 to 8, 19, 23, 27). These influences may be beneficial in so far as they promote vigour and eliminate some stresses (e.g. drought). In other cases, practices such as monocrop cultivation may obscure the initial effects of stress. Thus, selective felling of mature trees coupled with sanitation felling in forestry removes the individuals most likely to succumb to pollution and climatic stresses. Effects of stress upon natural regeneration are often obscured by the widespread practices of reseeding and replacement planting of seedling trees.

36. Management involves the selection and breeding of genotypes in agriculture, horticulture, forestry and fisheries (8). This has occurred over many centuries with the deliberate objective of improving productivity by direct (e.g. enhanced growth, fecundity, etc.) and indirect (e.g. improved establishment, disease resistance, etc.) means. Relatively little work has been done with the intention of systematically improving tolerance to air pollution although notable exceptions do exist in O_3-tolerant tobacco cultivars, and in SO_2- and metal-tolerant grasses (9). Screening experiments (e.g. National Crop Loss Assessment Network of the United States of America (NCLAN) (39)) also indicate that certain cultivars of commercial crops are much more sensitive than others to O_3 pollution. While the benefit of such information to farmers in polluted areas is obvious, there are indications that tolerance incurs a minor penalty when tolerant plants are grown in non-polluted environments.

37. It is much harder to judge the impact on dose-response functions of inadvertent selection of tolerant crops and even of individual cultivars by farmers, foresters and breeders working for many years in polluted areas. However, it is probably a pervasive and significant impact, judging by observations of changes in grasslands where exposure to ambient atmospheric pollution has resulted in the natural selection of tolerant strains and clones within a few decades (16). This natural process is probably slower in longer-lived plants such as trees, and in complex ecosystems such as forests and lakes, than in intensively managed agricultural arable systems.

III. DIRECT EFFECTS ON VEGETATION

38. In this section, available information on cause-effect and dose-response relationships between ambient air pollution and direct damage to vegetation will be examined. Numerous source documents have been consulted including many not listed in the references.

A. CELLULAR, BIOCHEMICAL AND PHYSIOLOGICAL EFFECTS

Gaseous pollutants

39. Ozone is a highly reactive pollutant affecting plasma and organelle membranes and consequently cell permeability. The precise mode of action is uncertain but O_3 is known to react with intracellular components such as glutathione and functional groups of proteins. It inhibits the formation of ATP in phosphorylation (53, 54). The partitioning of biomass between parts of the plant is altered, and the translocation of photosynthate to root systems is reduced. The reduction in root biomass and function in turn reduces water and nutrient uptake from the soil and supplies to the shoot. There is, therefore, a clear physiological mechanism for interaction between O_3 and water stress (56). High levels of O_3 generally cause stomatal closure which reduces further uptake of O_3 and inhibits loss of water vapour (8, 18).

40. SO_2 is very soluble in water where it dissolves to form sulphuric acid which establishes equilibria with its dissociation products, bisulphite (HSO_3^-) and sulphite (SO_3^{2-}) ions according to the pH of the solution. Sulphite may be oxidized to sulphate (SO_4^{2-}) and then metabolized by the sulphate reduction pathway (55). Some of these dissociation products are normal plant metabolites and consequently may be stored. In some cases they act as nutrients (e.g. to rectify soil sulphur deficiencies). Damage occurs when exposure to SO_2 results in rates of production of intermediate oxidation products which exceed the ability of the plant to incorporate the additional sulphur by the sulphate reduction pathway. High concentrations of SO_2 also cause stomatal closure with the same consequences as described for O_3 (8, 10, 16). However, at low concentrations (around 10 ppbV) and relative humidities (< 40 per cent), stomatal opening may be enhanced, especially when the plant is stressed by drought.

41. Small excesses of SO_2 oxidation products are associated with temporary metabolic disturbances in net photosynthesis, translocation, ATP production, etc., which are reversible (16, 57). Fatty acid metabolism is especially sensitive to SO_2 pollution (26). Large excesses result in massive accumulation of oxidation products, cell necrosis and early abscission of the leaf. The increased acidity may affect stomal reactions. There is evidence of inhibition of root growth, and of reduced water and nutrient uptake in SO_2-stressed plants (8, 10).

42. Oxides of nitrogen (primarily NO_2 and NO) are not very soluble in water. However, the uptake of NO into cells appears to be sustained by oxidation to NO_2 and by various reactions in the extracellular water which enhance the solubility of NO_2 (59). Once in the intracellular fluids, the NO_2 appears to enhance nitrate reductase activity which converts NO_3^- to NO_2 (58, 60). This appears to surpass the capacity of the nitrite reductase pathway by which NO_2 is converted to NH_4, resulting in an accumulation of NO_2. However, there is evidence to suggest that nitrite reductase activity may also be stimulated by NO_x pollution during periods of active growth (16). This may be related to the nutritional role of nitrogen and may account for reports of stimulated growth of plants exposed to relatively low concentrations of NO_2. Combinations of NO_2 and SO_2 at relatively low concentrations have the reverse effect of inhibiting nitrite reductase activity which may account in part for the reports of synergistic effects of the two pollutants (61). High concentrations of NO_2 appear to affect stromal reactions, and inhibit root growth and function.

43. The above review has concentrated on the ubiquitous gaseous pollutants O_3, SO_2 and NO_x. There are some noticeable similarities of effect which probably account for interactive phenomena associated with mixtures of the pollutants. An extensive body of literature exists on the effects of HF which is a potent inhibitor of many functional enzymes. The literature on other gases is less extensive but the information on peroxyacetyl nitrate (PAN) indicates that it is a transitory pollutant of high phytotoxicity, peak concentrations usually being associated with incidents of O_3 pollution (18).

44. Gaseous pollutants, in general, inhibit plant growth and development by one or more of the following actions:

(a) Inhibition of functional groups of enzymes;

(b)   Overloading of metabolic pathways for transforming or "detoxifying" potential toxicants; and

(c)   Damaging the integrity of the plasma and the cellular membranes.

The precise modes of toxicity vary and may be complicated by plants using certain pollutant oxidation products as nutrients under some conditions. The net effect of pollution at ambient conditions, however, appears to be a combination of inhibited cellular productivity, inhibited root growth and function, and reduced translocation and biomass production.

Non-gaseous pollutants

45.   The direct effects of non-gaseous pollutants in dry and wet deposition at ambient concentrations are less easy to ascertain (8, 10, 62).   Despite extensive experimentation few significant biochemical and physiological responses have been detected in vegetation exposed to present levels of rainfall acidity. There are, however, indications that:

(a)   Foliar leaching is influenced (see paragraph 22);

(b)   Pollen germination of some trees may be inhibited by rain at pH below 4 (10, 65);

(c)   Pollen tube elongation of some trees may not occur at pH below 4 (10, 64);

(d)   Stimulation or inhibition of the germination of spores of some pathogenic fungi may occur in the range pH 3.0 to 4.5 (8, 10, 63);

(e)   Microbial enzyme activity and growth of mosses in upland bogs may be reduced at current levels of acidity in rain (18, 19). The most acidic throughfall may even kill mosses (98);

(f)   Erosion of the cuticular wax of conifer needles may be enhanced in areas where trees are exposed to relatively high ambient levels of SO_2 and acidity in rain (17, 67).

The consequences of these effects may not be readily observed in the field.

46.   There is some evidence of direct toxicity of heavy metals deposited in very high concentrations as dusts on foliage already exposed to high ambient SO_2 pollution, but these effects are confined to industrial areas (9, 10). Some foliar uptake of metals may occur under normal conditions thereby enhancing total uptake via the roots in polluted soils. Unfortunately, the data on the tissue concentrations at which absorbed metals become toxic to plant cellular systems are extremely limited (10). Responses are complicated by the toxic effect on root systems of metals in solid solution, and by the nutritional roles of some metals in the context of indirect effects upon complex systems (see para. 78).

B.   EFFECTS ON WHOLE PLANTS AND CROPS

Relative sensitivities

47.   Many lists have been published of relative sensitivities of species and cultivars to SO_2, O_3, NO_x, HF and other pollutants. Such lists must be treated with extreme caution because relative sensitivities change with: (a) increasing or decreasing dose;   (b) various mixtures of pollutants;   and (c) variations in other stresses (e.g. drought, temperature) (8,9).   Nevertheless, some generalizations are possible.

48.   Generally, plants are less sensitive to NO_2 or NO than they are to SO_2 or O_3 or even HF. Legumes (e.g. Medicago, Pisum, Trifolium) are fairly consistent in their sensitivity to O_3. Winter wheat is possibly more sensitive to O_3 than barley; Ponderosa Pine and White Oak are among the more sensitive tree species. As far as SO_2 is concerned, Norway spruce is relatively more sensitive to large concentrations whereas Sitka spruce is the one more sensitive to small concentrations. While data for NO_2 are sparse, Sphagnum, some members of the Solanaceae and some woody species are reported to be sensitive to it (16). In all three cases there is considerable variation in sensitivity to the pollutants within a genus and between the provenances, cultivars and clones of an individual species. Cultivars of tobacco and lettuce, for instance, vary enormously in their sensitivities to O_3 pollution (18).

49.   In general, mixtures of SO_2 and O_3 are either "additive" or "antagonistic" in concentrations which individually would be sufficient to cause some inhibition of growth. At lower concentrations, however, a "synergistic" response may occur. Mixtures of SO_2 and NO_2 may be "stimulatory" at very low concentrations (see Figure 5) and "additive" or "synergistic" at higher concentrations.   This may apply to mixtures of SO_2, NO_2 and O_3. These responses may vary in magnitude according to growing conditions and the influences of abiotic stresses (8, 10, 16, 20).

Critical levels

50.   Despite the many reservations already stated, the following generalizations may be made

for sensitive plant species and cultivars. Concentrations are yearly average mean daily values unless otherwise stated (see also Figures 1 and 2):

(a) Long-term stimulation of growth has been observed only in agricultural crops deficient in sulphur and/or nitrogen when the atmospheric concentrations of SO_2 and/or NO_2 were equal to or less than 10 to 20 ppbV.

(b) Minor inhibition of growth may start when:

(i) SO_2 concentrations exceed 10 to 20 ppbV, especially when plants are under stress and growing slowly;

(ii) NO_2 concentrations exceed $\geq$ 50 ppbV NO_2;

(iii) Ozone concentrations exceed $\geq$ 100 ppbV per hr. day^{-1}; or ozone exceeds $\geq$ 25 ppbV per 24 hr. day^{-1};

(iv) $\geq$ 15 ppbV SO_2 plus 15 ppbV NO_2 plus 30 ppbV O_3 (16).

(c) Significant inhibition of growth, with or without foliar injury, may occur at:

(i) $\geq$ 60 ppbV SO_2;

(ii) $\geq$ 50 ppbV SO_2 plus 50 ppbV NO_2;

(iii) $\geq$ 80 ppbV O_3 (especially if PAN is present); and

(iv) $\geq$ 40 ppbV SO_2 plus 40 ppbV NO_2 plus 40 ppbV O_3.

51. As concentrations or doses increase above these critical levels, there will be a proportionally greater reduction in growth and an increase in foliar damage to sensitive species and cultivars. A wider range of less sensitive species will also begin to be affected. As described in section I, the threshold limits correspond to ambient pollution levels common in many areas of Europe and North America. The definition of relevant dose-response functions and quantification of losses under field conditions, however, are not easy tasks.

Dose-response functions

52. Some useful data are beginning to emerge not only from experiments in laboratories and glasshouses (greenhouses) but especially in the field (8, 10, 16 to 18, 34). Roberts pooled experimental and field data from experiments with rye grass exposed primarily to SO_2 pollution; he derived the following equation:

$$y = + 7.33 - 0.21 \ SO_2$$

where y = percentage yield loss, and SO_2 = concentration of gas in ppbV. This equation indicates a threshold of $\geq$ 40 ppbV SO_2 (68). Mansfield and Freer-Smith, however, separated many of the same data into two groups: one with SO_2 added to prefiltered or clean air, and the other with SO_2 added to ambient air (59). The first group complied with Robert's basic function; they indicated possible stimulation of growth at 20 to 40 ppbV SO_2 and included data for the Cowling et al. experiment (69) on fast-growing grasses in S-deficient soils. The second group, however, indicated appreciable decreases in growth at ambient levels of air pollution, especially when other stresses including additional pollutants were present (61, 70, 71). Other experiments have shown the same distinction between clean or filtered and ambient air exposures (8).

53. One open-top chamber experiment involved the addition of O_3 to prefiltered air which consistently enhanced the growth of snap beans compared with beans exposed to the same total of O_3 in ambient air (72). Another experiment used 18-22 ppbV SO_2 and 1.0 ppbV HF in prefiltered and unfiltered air, the unfiltered treatment resulting in 32 per cent more yield loss in spring barley cultivars (73). Heck et al. recorded yield losses in peanut (-18 per cent), lettuce (-12 per cent), soyabean (-11 per cent), turnip (-7 per cent) and kidney bean (-2 per cent) attributable to a seasonal exposure of 25 ppbV O_3 per 7-hour day^{-1} mean in ambient air. The experiment was part of NCLAN (13) and obtained significant linear regressions between exposure to O_3 and yield losses in these crops in 1981 in the United States. The regression varied from:

$$y = 1065.7 - 5.678x \text{ for lettuce}$$
(cv. **Empire**); and

$$y = 16.8 - 0.031x \text{ for kidney bean}$$
(cv.**California Light Red**);

representing the most sensitive and tolerant extremes, respectively, where y = yield and x = seasonal 7-hour day^{-1} O_3 concentrations in ppbV.

54. The above functions all assume a linear dose-response function. However, another desk study pooled dose-response data for foliar visible injury and O_3 pollution, and derived:

$$y = 2.14 + 0.0055x$$

where y = percentage foliar injury and x = $(\log^{10}$ pphm O_3) + $(\log^{10}$ hr x 10) which approximates a quadratic or sigmoidal dose-response function (9). It is possible, therefore, that the data described above reflect:

(a) the variability of the dose-response function and

(b) the inherent instability of the lower ranges of dose-response relationships.

Both issues were tested with NCLAN experiments which reveal a close interaction between O_3 dose/plant yield response and moisture stress (Figure 3) (33). In view of the latter data, it is important to assume a dose-response function of considerable uncertainty (Figure 5) for SO_2, NO_2 and O_3 alone and in various combinations. Much more research is needed in this area before defining the direct dose-response functions of major agricultural and forestry species subjected to ambient mixtures of gaseous and other pollutants. The general mode (Figure 5) will have to suffice until such data are available. This issue will be taken up later with special reference to forest injury, damage and decline.

C. DIRECT INTERACTIONS WITH BIOTIC STRESSES

55. Plants in a polluted atmosphere are also exposed to attack by pests and parasites. There is considerable evidence of both positive and negative but direct effects upon such organisms (8, 10, 63). Most primary parasites, especially obligate parasitic fungi and bacteria are inhibited by gaseous and particulate pollutants including acidic materials. Spore germination and infection are reduced at around 40 ppbV SO_2 and by leaf surface moisture pH values of < 4.0. However, lower levels of pollution may stimulate germination and infection in some cases. These responses represent a direct relation from pollutant to parasite.

56. In other cases, the cause-effect relationship appears to involve some modifications by the pollutant of the hosts's resistance to a pathogen or parasite (10, 63). This appears to be especially common with plants exposed to O_3 and oxidant pollutants, there being many examples of both stimulation and inhibition of obligate parasitic fungal, bacterial and viral parasites.

57. Facultative foliar and stem pathogens generally seem less sensitive to air pollution than obligate pathogens and more able to exploit opportunities for invading plants weakened by exposure to air pollution, especially when physical injuries offer avenues for infection. The position of soil-borne pathogens is somewhat similar in so far as anything which reduces root vigour increases the risks of infection (10, 74). However, reductions in soil pH below 4.0 appear to reduce the activities of some soil-borne pathogens but may stimulate invasion by others (e.g. **Armillaria sp.**) (74, 75).

58. There is increasing evidence that invertebrate pests are affected by air pollution (4, 8, 10, 76 to 79). Some toxicants (e.g. HF, heavy metals) reduce the fecundity and fertility of aphids and other pests. In other cases, however, pollution appears to increase populations of certain pests and their damage to foliage, etc. Three processes may be involved: (a) modification of host defense mechanisms; (b) enhancement of host palatability to pest; and (c) selective inhibition of predators upon pest organisms, any of which can result in population explosions of serious proportions, especially in monocrop systems exposed to air pollution.

IV. COMPLEX EFFECTS ON VEGETATION AND SOILS

59.   This section examines cause-effect relationships and dose-response functions associated with: (a) air pollution-induced changes in soils; and (b) combined direct and indirect effects in complex systems, notably forests.

A. SOILS

60.   An ecosystem may be considered in simple terms as a box model with inputs and outputs, and with subcompartments (Figure 6). Numerous exchanges take place between the sub-compartments and the outside world.   A system in its climax state (paragraph 31) will minimize outflows and maximize cycling of energy, nutrients, etc. within the system (10, 80). The introduction of air pollution adds materials which create imbalances in the system and stimulate outflows, especially from the soil "subcompartment".   The constituents of those outflows are determined largely by the nature of soil and bedrock.

Bedrock

61.   The original nature of a bedrock determines in a particular location:

(a) The nature of the overlying soil created by weathering of the bedrock and subsequent biological, physical and chemical processes;

(b)   The physical and chemical characteristics of surface run-off and soil drainage waters, and hence of surface water and groundwater;

(c)   The optimum natural complex of vegetation which will develop and be maintained or "carried" on that soil; and

(d) The ability of the soils and ecosystems to cope with various forms of pollution.

62.   Soils and bedrock vary enormously in their physical and chemical characteristics. Thus, those most able to cope with metal pollutants are generally characterized by a high content of organic matter in the soil and the ability to immobilize metals in non-toxic organic complexes.   Conversely those areas most at risk from acidification (23) generally:

(a)   Are acidic already and contain negligible glacial drift;

(b)   Contain little or no carbonate minerals naturally, or because they have been removed by weathering;

(c)   Contain little readily weathered and reactive minerals (e.g. feldspars, mica); and

(d)   Support only shallow soils and/or exhibit rapid through-flow of precipitation so that the residence time of water is low.

Such areas may be found on many sands, gravels, granites and other bedrocks which are base-poor and slowly weathered.

Figure 6

Main components of a simple box model of ecosystem/soil bedrock relations with atmospheric and aquatic systems (10)

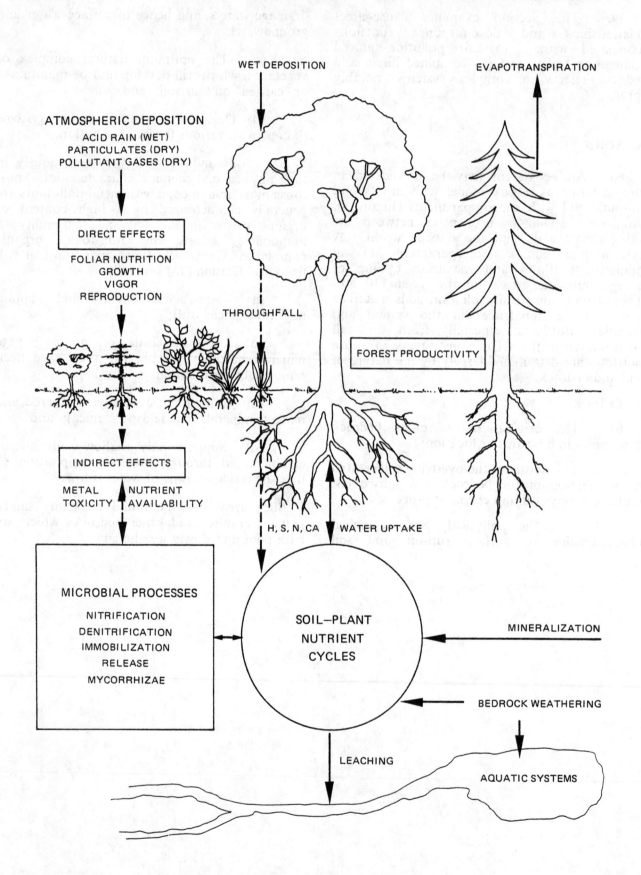

Soil acidification and cation leaching

63.   Interpretation of cause-effect relationships is hampered by the complexity of the soil and its processes, by gaps in our knowledge of biogeochemical processes, and by wide variations in soil type, geology and land-use practices. The time scales of significant changes in some soil properties partly attributable to acidic deposition may cover periods of 20 to 50 years. Nevertheless, acidification is known to be a relatively slow process, rain being naturally mildly acidic at $\geq$ pH 5.0. The acidifying potential of atmospheric deposition is more closely related, however, to the ratio of H^+ to NH_4^+, Ca^{2+} and Mg^{2+} than to the H^+ concentration of the deposition per se (6). The uptake of cations by plant roots is generally higher than that of anions, cations being released only when leaves and other material return to the soil and are decomposed. To compensate the ionic imbalance in uptake, roots release H^+. There is thus an imbalance of H^+ in the soil which may be neutralized by base cations derived by weathering of the bedrock.

64. In most plant systems, however, the plant biomass grows and litter accumulates in the surface layers of the soil as the plants age. The accumulation of minerals is faster than their release by decomposition and mineralization. There will thus be net soil acidification (6, 10, 80, 81, 82). This process is most striking in upland conifer plantations, before canopy closure. Potassium, for instance, virtually disappears from soil drainage waters (19). Eventually, the system reaches maturity, and soil acidification slows down then stabilizes.

65.   Soil acidification is not necessarily synonymous with cation leaching (10). Leaching without acidification may occur where the weathering of a base-rich bedrock is rapid enough to resupply the system with cations, to retain $CaCO_3$ in the soil, and to hold the pH of soil solution at between 6 and 8. Once the $CaCO_3$ disappears, or is in short supply, then the weathering of silicate minerals and cation exchange processes take control of soil buffering (6, 10, 23). When the pool of exchangeable Ca, Mg, K^+ and Na^+ begins to be depleted, leaching of exchangeable H^+ and aluminium starts to increase. The aluminium pool is almost unlimited and leaching outputs will remain high as acidification continues.

66. Soil with a high capacity to immobilize sulphate and nitrite can receive large amounts of acid without being markedly acidified.

67.   Soils at risk from acid deposition are those of intermediate pH and fertility but poor in easily weathered base minerals (6, 10, 23). Such soils will rapidly be depleted of base cations especially if they are thin, freely drained, and subject to pulses of H^+. Acid mineral soils may also be at risk because they have small reserves of nutrients and respond with increased aluminium release. The combination of natural acidification, acid deposition and afforestation may have pushed both types of soils along these predicted roads (10). The precise contribution of acid deposition to soil acidification and degradation is, however, not easy to ascertain.

68.   The deposition rate (section I) of H^+ ranges from 0.2 to 0.6 keq ha^{-1} yr^{-1} in remote areas, through 1.0 to 2.0 keq in much of central Europe and North America, to 3.0 to 4.0 keq in more polluted areas. Potential acid deposition (H^+ and NH_4^+) can in some instances reach values exceeding 6 keq ha^{-1} yr^{-1}. Values of 1.0 to 4.0 keq ha^{-1} yr^{-1} exceed the capacity of normal weathering processes (est. 0.2 to 2.0 keq H^+ ha^{-1} yr^{-1}) in many soils of Europe and North America. Some of the field data are contradictory but this is not surprising in view of the complex processes and long time scales involved. For instance, the exchangeable reserves of a deciduous forest in Tennessee (United States of America) were estimated at 50 to 70 years despite a doubling of internal acid production (83). Nevertheless, analyses of transects from main sulphur emission sources have shown a clear relationship in some instances between declining topsoil pH and distance from source. There are reports also of 0.5 to 1.0 pH unit declines over 25 to 50 years in certain forest soils in central Europe, North America and Scandinavian countries (4, 6, 10, 97). Tamm (97) has found that the acidification of mineral soil is caused by the deposition of strong acids. Part of the decrease in pH in the humus layer can be caused by biological processes.

Nitrogen cycling

69.   Emissions of NO_x and NH_3 result in increased deposition of nitrogen, which is a plant nutrient, as NH_4^+ and NO_3^-. Deposited NH_4^+, often in the form of ammonium sulphate and nitrate, is absorbed easily by vegetation, thereby enhancing growth. Similarly, NH_4^+ and NO_3^- in the soil are available for root uptake and plant

growth. Deposited nitrogen is accumulated in the organic layers of the topsoil and eventually stimulates nitrification by soil micro-organisms. Deposited ammonia is not very mobile in soil and a surplus can either be accumulated in topsoil or converted to nitrate by nitrifying bacteria, thus contributing to soil acidification. If nitrate is leached from soil, with base cations, increased acidification of the soil will occur (23).

70. Wet and dry deposition of mineral nitrogen compounds in remote "unpolluted" areas is estimated at 1.0 to 3.0 kg N ha^{-1} yr^{-1} in temperate coniferous forests, the yearly accumulation of nitrogen in tree biomass being 10 to 60 kg N ha^{-1} yr^{-1}. Experiments with fertilizers indicate an optimum input of 15 to 25 kg N ha^{-1} yr^{-1} for the growth of coniferous forest. Increased inputs of N may, however, increase top growth and raise susceptibility to stresses, such as drought, frost, wind and grazing damage. However, such responses may not occur where growth is limited by the availability of other nutrients (e.g. where PO_4^{3-}, Mg^+ are limiting factors (19)). In these instances, deposited nitrogen contributes to soil acidification.

71. In forests where atmospheric nitrogen inputs are excessive, enhanced leaching of nitrate might be expected especially if growth is limited by other factors. The threshold for enhanced leaching appears to be 5 to 15 kg ha^{-1} yr^{-1} deposited nitrogen. Experiments in North America indicate that 6.7 to 20.0 kg N produced leaching rates of 20 per cent and 60 per cent total input, respectively (45). Soil acidification and cation leaching are likely to be further aggravated where high deposition rates of H^+ and/or SO_2 and NH_4^+ and NO_3^- and/or NH_3 occur simultaneously.

Mobile ions

72. There is a close correlation between the formation and leaching of NO_3^- and the dissolution and leaching of aluminium in soils. There is empirical evidence that the solubility of most metals is enhanced and that their biological availability is increased by higher soil acidities. A decrease of 0.1 to 0.2 pH units in temperate soils may be accompanied by a two-fold increase in soil solution. Concentrations of 5.0 to 20 mg aluminium l^{-1} are reported frequently in the waters of soils exposed to acid deposition in Europe (32, 84). Also, most of the aluminium in lakes in such areas appears to originate from soils within the lake catchments (24, 25).

73. Such simple statements belie problems with the complexity of soil and the close relation between dissolved aluminium and dissolved organic matter. Aluminium is often singularly immobile in topsoils with a high organic content. This relationship is not closely dependant on pH although aluminium may become increasingly mobile as pH declines in the lower horizons of the same soils. The pool of aluminium which is leached varies according to soil type. The close link between aluminium and nitrate leaching suggests also that the breakdown of immobilized organic aluminium complexes may be closely related to nitrification and enhanced where NH_4^+ and NO_3^- deposition is high (6).

74. Leaching of SO_4^{2-}, on the other hand, appears to be closely related to atmospheric inputs of sulphur. As such inputs increase, most systems move rapidly to a new equilibrium between input and outputs in soils and catchments (2, 23, 25). K^+ is also highly mobile but rapidly taken up by vegetation, especially trees, so that little of it leaches into afforested catchment waters (19).

75. The budgets of metals in forest soils are extremely complex (6, 10). Pb, Cu and Fe tend to be retained in most forest soils but Ca, Mg, Cd, Zn and Ni tend to be released into groundwater, the concentrations increasing with rising acidity. Not surprisingly, plants growing in acid soils tend to have higher tissue concentrations of Cd, Pb, Zn and also of Mn and Ru but lower concentrations of Ca and Mg than the same or related species in more alkaline soils.

76. Unfortunately, there is very limited information on the soil and tissue concentrations at which metals become toxic to plants, especially for forest species in acid soils (10). Furthermore, the toxicological interactions between metals and nutrients in soils are exceedingly complex. For instance, a metal essential for growth could be available for uptake in soil but the plant may still exhibit a deficiency because uptake is impaired by toxic or deficiency combinations of other elements. Nevertheless, some generalizations are possible.

77. Experiments with tree seedlings indicate that the toxic threshold for most species is 1 to 160 mg Al^{3+} l^{-1}. There seems to be only a slight toxic or antagonistic effect if the calcium/aluminium ratio is larger than 1.0 and/or if the content of organic matter is high. The experimental results indicate that the concentration of aluminium found in soil water in many areas of Europe affects the growth and chemical composition of roots in mineral soils whereas in the humus layer most aluminium is organically complex and therefore not very toxic. It has also been postulated that high concentrations of aluminium and other ions may

inhibit root and mycorrhiza formation (32). High NO_3^- concentrations have much the same effect. The situation is further complicated in cases where exposure to gaseous pollution also inhibits root development (section III above). These mechanisms may also explain in part the reported foliar deficiencies of Mg in trees exposed to acid deposition and soil acidification in Europe and North America (4, 10). However, the Mg deficiency syndrome may include other depletions associated with: (a) gaseous damage to chloroplasts (section III); (b) leaching of Mg from foliage by run-off of acidic wet precipitation; and (c) normal depletion processes associated with senescence and abscission. In the Netherlands a fourth mechanism has been recognized, namely the competitive uptake of ammonia.

78. The fertility of the soil depends upon the activities of soil organisms, many of which are known to be affected by acidification and/or by increases in the availability of nutrients and metals in the soils (6, 10). The effects of increased nitrogen deposition have already been discussed. In general, the groups of organisms most likely to be inhibited by acidification and/or increased concentrations of metals include macrofungi, earthworms, beetles and collembolans - all important in the decomposition of organic matter and the recycling of nutrients. However, many organisms can tolerate very high concentrations of seemingly toxic metals under acid conditions, some such as woodlice being exceedingly tolerant (6, 85). Such tolerance appears in most cases to be at the expense of productivity which means that the niches vacated by sensitive organisms are never fully filled by tolerant organisms, thus producing a less fertile and less productive or less "energetic" system.

Critical loads

79. The preceding discussion provides some useful background information by which to examine later combined effects on forests and eventually indirect effects on surface water and groundwaters. Soil is so variable and complex that it is impossible at present to make sound generalizations. However, the following guidelines may be of use:

(a) **Hydrogen inputs:**

(i) Any extraneous acidic input is likely to increase the rate of acidification in forest soils. The capacity of most forest soils to neutralize acid input (critical loads) falls in the range of 0.1-1 keq. ha^{-1} yr^{-1} for most non-calcarous soils (96). The alkalinity production of the susceptible soils in Scandinavia usually is in the lower part of the range. These critical loads are in many areas much lower than the actual rates of deposition; as validation, pH changes in the soil of 0.5 to 1.5 have been recorded over 30 to 50 years in many studies in Scandinavia and central Europe.

(ii) Increased soil acidity, for instance by 0.3 to 0.5 pH units, is known to have a number of negative effects on soil processes and plant productivity.

(iii) Decreases in the amounts of acidic inputs will slow the rate of soil acidification but are unlikely to reverse it.

(b) **Nitrogen deposition** : It is not possible on the basis of present knowledge to draw firm conclusions on critical loads. However, from various experimental studies and field observations it can be concluded that at deposition rates of 10 to 30 kg N ha^{-1} yr^{-1}, changes take place within the soil and the vegetation owing to nitrogen saturation of vegetation (16, 97). These changes include enhanced leaching of nitrate from the soil, shifts in vegetation, enhanced frost sensitivity and reduced photosynthesis.

B. COMBINED EFFECTS IN FOREST SYSTEMS

80. In section II the concepts of cause-effect relationships and dose-response functions were set in their ecological context with special reference to forests. The various direct and indirect effects of air pollution that may occur under ambient conditions have already been examined. The complex problems of defining cause-effect relationships and dose-response functions over the extended time scales relevant to forests are well known. This section deals with the evidence of forest damage and the extent to which it can be attributed quantitatively to the effects of air pollution.

81. Trees grow for decades, the optimum age for harvesting ranging from 40 to 120 or more years (5, 10). Forests are, therefore, exposed to repeated doses of pollutants in all possible combinations and in all possible forms of dry and wet deposition. The nature of the dose varies with location, season and year; trees thus suffer both the shorter-term direct effects (e.g. of SO_2, O_3, etc.) described in section III, and the longer-term indirect effects of soil modification described above under section IV.A. The combined result of these effects is modified extensively by other short- and long-term stresses

(e.g. diseases, climatic change) and management practices. One would expect, nonetheless, that the effects of these combined stresses might be manifest first in trees which are: (a) mature; (b) exposed to extremes of weather; and (c) at or near the limits of their altitudinal and latitudinal distributions. One might also expect that widespread stress would result in general

Assessment of damage

Yield losses

82. Some assessments are based largely on evidence from fumigation experiments (c.f. section III for SO_2, NO_x and O_3). Air pollutants can inhibit growth and, therefore, reduce productivity in forests, with or without visible injury. Some allowance may also be made for indirect effects of an unspecified nature but there is no discrimination between species of trees. The final assessment of yield loss is based on the informed judgements of experts familiar with the scientific literature and with the forest industry and ambient levels of air pollution. Into this category fall published and unpublished national assessments of losses in productivity of forests and agriculture in Canada (29), the Netherlands (15), the United Kingdom (14), and the countries of the European Economic Community (EEC) (36). They generally assume a range of 1 per cent to 10 per cent yield loss. The primary causal agent may be specified (e.g. O_3 in the Netherlands) or generalized as "air pollution". Field evidence to support these estimates had been largely lacking until the 1980s when field surveys and research began to offer both circumstantial evidence of possible damage and a possible new approach to damage assessment.

Foliar damage

83. Concern was first expressed in the early 1980s about the more widespread distribution of damage which could not be readily explained (2, 4, 5, 10). Symptoms included needle/leaf chlorosis and premature abscission, shoot dieback, crown damage, and epiphytic shoot production followed by the death of the most mature specimens affected. Some species were affected more than others. National programmes were mounted in Europe from 1983 onwards in order to survey the extent and severity of injuries. Various survey techniques were used in each country so that the results are not strictly comparable. In addition, most of the surveys measured the percentage of damaged trees. Extrapolations to the area of damaged forest may

reductions of tree development and growth, with or without symptoms of visible injury to foliage. Proof of such phenomena is, however, hard to obtain. Consequently most assessments of damage are based on national survey and monitoring programmes using visible symptoms in addition to measurement of stem increment and foliar analysis.

therefore be misleading as trees with varying degrees of damage are often found next to perfectly healthy trees. Nevertheless, broad trends do appear (5).

84. For 10 countries in Europe, trees equivalent to some 6.1 million hectares of forest are now estimated to be affected. Of these, 1.7 million hectares are moderately damaged while 0.23 million hectares are dying or dead (Table 1). Data from the Federal Republic of Germany, Luxembourg and Switzerland suggest a doubling of damage between 1983 and 1984, although this may be owing in part to an intensification and improvement of survey techniques. The Swiss data show an increase in damage with altitude. Damage appears to be most severe and was increasing in 1984 in a contiguous zone touching: Austria; Belgium; Czechoslovakia; France; German Democratic Republic; Germany, Federal Republic of; Hungary; Luxembourg; the Netherlands; Poland; and Switzerland - exceeding 50 per cent of total trees in some areas (5). Some damage is reported in other countries but it is generally considered minor, and decreasing towards the northern and western limits of Europe. Various studies suggest that similar damage is most common in the eastern and north-eastern regions of North America. The extent to which the early stages of damage are reversible is unknown. Available surveys for 1985 indicate that the increase in damage between 1984 and 1985 was minor, possibly because of the wet summer of that year.

85. The risks of using non-specific symptoms to describe a cause-effect relationship or dose-response function have already been mentioned in section II. These are well-illustrated by the purported causes of foliar damage attributed to various combinations of air and soil pollution, abiotic and biotic stresses. Air pollution in one form or another is generally cited as an important factor. Given the present state of knowledge, the results of these surveys are best regarded as a general assessment of the health of forests.

Radial growth

86. Further possible evidence of sustained damage comes from studies of tree-ring growth increment in Europe and North America (4, 10, 86). Reductions in radial growth of between 10 per cent and 25 per cent per annum appear to have started in some, but not all species between 10 and 30 years ago in the same areas as those exhibiting foliar injury. Mature trees are most affected. In contrast, there are instances of moderate defoliation without any reduction of increments. Note that the onsets of foliar damage and increment reductions are not contemporaneous, there being an "unexplained" gap of at least 10 years. Nevertheless, both are considered symptomatic of forest damage and decline, and possibly caused by pollution.

Climatic changes

87. The possibility of interaction between air pollution and both short-term and long-term climatic changes cannot be ignored. Baumgartner (52) (cf. para. 33) summarizes the possible effects on temperate forests of seemingly minor changes in climate. These changes imply differential responses between species especially in montane areas. Other researchers have been unable to relate climatic changes to reductions in tree growth but can find an association with ambient levels of air pollution (92). Evidence of long-term climatic changes comes from palaeobotanical, palaeolimnological, glaciological, meteorological and oceanographic studies (27). A link between these changes and effects on trees is, however, not yet established. A recent World Meteorological Organization (WMO) conference on world climate (27) tentatively identified a number of long-term changes although validation of some of them is still open to debate. These include:

(a) A warming of the western hemisphere between 1880 and the 1970s peaking in 1939/40;

(b) A cooling phase from the 1950s to the early 1980s during which the annual average surface temperature of the earth declined by 0.6 to 1.0°C and the annual average sea surface temperature declined by 0.1 to 0.2°C per decade; reverse trends of smaller magnitude occurred in the southern hemisphere;

(c) From 1954, winters became cooler with great monthly extremes (e.g. frosts) in spring and autumn;

(d) There may have been a reduction in mean annual rainfall in some areas of Europe and North America within the period 1880 to 1960; and

(e) The cooling trend levelled off in the early 1980s and according to the latest consensus (87), may be followed by a warming period of unprecedented speed and magnitude caused by the "greenhouse" effect of rising CO_2 and other emissions to the atmosphere.

88. The possibility of an interaction between climatic change, various forms of air pollution and other stresses cannot be ignored (10).

Dose-response functions

89. Manion has provided a conceptual framework for considering the roles of different stresses in forest decline (Figure 7) (93). The framework accommodates all of the possible direct and indirect effects of pollution described in section III. Despite the considerable survey and research work carried out to date (4, 5, 10, 11), it is still not possible to use this framework fully nor to identify precisely the cause-effect and dose-response relationships for climatic and pollution stresses involved in forest decline. For instance, the latest studies indicate considerable uncertainty about the yield losses and economic damage resulting from forest decline in Europe and North America. The crude estimates range from 1 per cent to 10 per cent per year yield loss, losses being highest in the contiguous area of central Europe and in the north-eastern part of North America. The impact of decline in terms of sanitary fellings of damaged mature trees, and the consequences for the forest industries is even more controversial (5).

90. How much of this decline is due to air pollution alone remains pure speculation. It will be resolved only by co-ordinated surveys, monitoring and research. The International Co-operative Programme for the Assessment and Monitoring of Air Pollution Effects on Forests (1) aims to integrate current national surveys and couple them with detailed studies of foliar damage, tree growth, analyses of tissue and soil characteristics at selected sites. At least some of these sites should also include detailed monitoring of air pollutant deposition, and of meteorological parameters to help unravel the interaction of climate and pollution on soils and tree growth. Despite the inherent technical problems, some attempt at experimental exposure of trees to ambient levels of pollution along the lines of the United States National Crop Loss Assessment Network (NCLAN) (13, 39) and the proposed EEC programme (94) for agricultural crops is essential if Koch's postulates are to be observed (cf. paras.25 and 26).

Figure 7

Some categories of factors influencing declines of forest trees (modified after Manion (93))

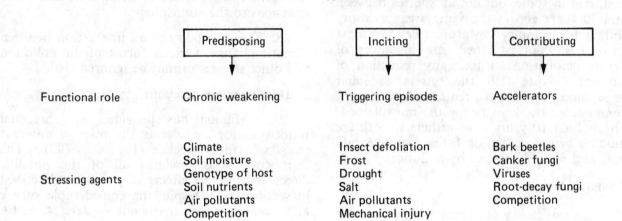

Types of influencing factors

	Predisposing	Inciting	Contributing
Functional role	Chronic weakening	Triggering episodes	Accelerators
Stressing agents	Climate Soil moisture Genotype of host Soil nutrients Air pollutants Competition	Insect defoliation Frost Drought Salt Air pollutants Mechanical injury	Bark beetles Canker fungi Viruses Root-decay fungi Competition

91. The general consensus is that the effects of air pollution should be considered in conjunction with other stresses, e.g. climatic change.

(a) For exposures to gaseous pollution (notably SO_2, NO_x and O_3), the generalized thresholds and dose-response functions were described in sections III and IV. Sensitivities and yield losses will increase with: (i) altitude; (ii) temperature stress; and (iii) moisture stress.

(b) Forests exposed to atmospheric deposition on soils of low base cation content and slowly weathering bedrocks are apparently at special risk from long-term soil acidification. Gradual reductions in soil pH and increases in the leaching of aluminium, SO_4^2, NO_3^-, etc. may be used as indicators of system degradation.

Some economic assessments in agriculture and forestry

92. From preceding discussions, it will be appreciated that economic assessments of damage are based on very limited data and imply a large measure of uncertainty. Nevertheless, the results are instructive.

93. The official Netherlands estimate of annual damage to agriculture stands at about 5 per cent of total yield, three-quarters being due to O_3 pollution. The benefits to consumers of reducing air pollution by 25 per cent were estimated at 250 million guilders (15). Comparable estimates of losses for other EEC countries per annum range from $US 900 million to 1,100 million and in the Federal Republic of Germany for forestry and agriculture, DM 1,000 million per year (36).

94. The United States NCLAN programme has provided hard data on damage, mainly that owing to O_3 pollution, for a wide range of agricultural crops and cultivars (13). Sensitivities varied among cultivars: annual yield losses ranged from 2 per cent to 23 per cent (39). The economic consequences are complex but various analyses indicate that a variation of + 25 per cent in ambient levels of O_3 pollution will produce net per annum economic benefits of + $US 2.6 billion to agricultural producers and consumers (13). Estimates of this magnitude have emerged from a number of other American studies (13).

95. It is interesting to compare the above estimates with some made in the late 1970s for the economic costs to forest industries of climatic changes of the same order as those discussed in the Canadian study (Table 2) (95). Note that the cost (positive sign) to Canadian forest production of a +0.5°C rise in mean global surface temperature was estimated at $US 2,360 million (1974 prices). It is also important to remember that the economic losses attributed to changes in both ambient levels of air pollution and climate are well within the range of meteorological events (e.g. floods) and pest infestations.

V. CONCLUSIONS AND RECOMMENDATIONS

96. Over the past few decades the nature of the air pollution problem has extended beyond local high concentrations of relatively short duration to relatively low concentrations of pollutants distributed over large regions for long periods of time.

97. Whereas short-term (acute) damage is easy to recognize, longer-term (chronic) damage, now more widespread, is harder to recognize and quantify. For example, according to present understanding of effects on forests in central Europe, a large number of factors - both natural and man-made - may be involved in complex interactions.

98. Interactions occur between different pollutants in the air, between direct effects of gaseous pollutants and indirect effects of pollutants through the soil as well as between pollutants and natural factors such as climatic and edaphic conditions.

99. The direct effects of single pollutants on plants are relatively well understood and different critical levels have been identified. Minor inhibition to plant growth may begin when SO_2 exceeds 10 to 20 ppbV (25-50 ug m^{-3}) (annual mean value), NO_2 exceeds 50 ppbV (100 ug m^{-3}) (annual mean value) and O_3 exceeds 100 ppbV (200 ug m^{-3} for one hour (16).

100. The combined effects of various gaseous pollutants are less well studied. However, some results do exist, especially for crops. Minor inhibition of growth may commence in a mixture of 15 ppbV of SO_2, 15 ppbV of NO_2 and 30 ppbV of O_3. However, given the current state of knowledge, it is still difficult to establish dose-response functions which are generally applicable.

101. (a) The concentration of SO_2 typically ranges between 50-160 ug m^{-3} in urban areas in

Europe and North America. In rural areas the mean annual values are typically 5-50 ug m^{-3}.

(b) The concentration of NO_x is in the range of 40-100 ug m^{-3} in urban areas and 1-40 ug m^{-3} in rural areas, in certain parts of Europe and North America.

(c) The concentrations of ozone exceed 120 ug m^{-3} for seven hours during 5 to 20 days per year in urban and rural areas in large parts of Europe and North America, with even large numbers at high altitudes.

102. Soil acidification owing to deposition of air pollutants is most intense in slightly acid, poorly buffered sandy soil. Acid podzolic soils are less likely to undergo further acidification owing to deposition, but the content of nutrients in these soils is low and even minor changes may have negative effects on forest trees.

103. Soil acidification is a natural process and the extra input of acid from the atmosphere will increase the rate of acidification. The soil acidification component caused by wet and dry desposition of sulphur- and nitrogen-compounds will, in several European countries, also include ammonium sulphate. In many types of soils the anions of the strong acids are mobile and will be leached out along with equivalent amounts of cations. Ammonium may either be taken up by plants or nitrified, both processess producing acids. Decreases in the amount of acid inputs will decrease the rate of soil acidification but are unlikely to reverse it.

104. Recent studies have shown three- to more than ten-fold increases in soil acidity (pH decrease of 0.5 to 1 unit) over 25 to 50 years in certain types of soils in Europe and North America. Increased soil acidity has numerous consequences, but the main effects on forests and other vegetation may be nutrient deficiency, for example Mg, and the development of toxic conditions in the root zone owing to mobilization of elements such as aluminium.

105. Susceptible soils will be at risk when:

(a) The total deposition of acid exceeds the alkalinity production (primary weathering) in the soil, which means a critical load of 0.2-1.0 keq ha^{-1} yr^{-1} for most susceptible soils. The actual deposition is 0.2 keq ha^{-1} yr^{-1} in northern Scandinavia but exceeds 4 keq ha^{-1} in some parts of central Europe. Thus a significant decrease of acid deposition is needed to protect soil in the long term.

(b) Long-term deposition of nitrogen exceeds 10 to 20 kg N ha^{-1} yr^{-1}. In the high-productive sites 20-45 kg N ha^{-1} yr^{-1} may be acceptable. The actual deposition is 1 to 3 kg N ha^{-1} yr^{-1} in northern Scandinavia, reaching 20 to 60 kg N ha^{-1} yr^{-1} in central Europe.

106. The area of forest damaged in 1984 in 10 countries of Europe where surveys were conducted has been estimated to be equivalent to about 6 million hectares. This is equiavlent to about 24 per cent of the total forested area of these countries. Of this total area, approximatley 17 per cent was regarded as slightly damaged, approximately 7 per cent moderately damaged and less than 1 per cent as dying or dead.

107. Various analyses in the United States indicate that a decrease of about 25 per cent in ambient levels of 0_3 pollution will produce net per annum benefits of $US 2,500 million to agricultural producers.

108. The reason for the forest damage is not well understood, but it is believed to be caused by a number of factors of which pollution and climate are considered important. However, factors related to forest management may also apply. For example, where the wrong provenances were used or where trees were introduced beyond their natural area, the damage may be worse.

109. In order to guide the work on emissions control to reduce as much as possible damage to natural vegetation, forests and crops:

(a) Critical levels for air quality and/or deposition should be defined for different types of soils based on the present state of knowledge, and geographical areas exhibiting higher levels and loads should be mapped as a basis for assessing potential damage.

(b) More attention should be focused on the effects on NO_x, ozone and hydrocarbons on the environment.

110. It is recommended that workshops be held on setting critical loads for the response of forests and crops, beginning with sulphur and nitrogen compounds and ozone.

111. Models should be developed (using, inter alia, EMEP facilities) for the calculation of emission reductions necessary to comply with given critical loads.

112. Better use should be made of data from existing national and international research programmes and more research should be initiated in order to assess the effect of climatic changes and other environmental stresses as well as of mixtures of pollutants on vegetation.

TABLE 1

**Estimated area of visible damage to crowns attributed to air pollution
(data as of May 1985)**

	Light damage	Moderate damage	Dying/ dead	Percentage of exploitable closed forest area		
				Total area damaged		Dying/ dead
	(1 000 ha)				(%)	
Austria	240 *	80 *	10 *	330	10	0.3
Belgium	17 *	2 *	1 *	20	3	0.2
Czechoslovakia	514 *	123 *	54 *	691	16	1.2
France	86 *	11 *	3 *	100 *	1	-
Germany, Federal Republic of	2 424	1 163	111	3 698 *	50	1.6
Hungary	103 *	13 *	4 *	120	8	0.3
Luxembourg	25 *	5 *	- *	30 *	37	-
Netherlands	80 *	20 *	- *	100	34	-
Poland	419	199	36	654	8	0.4
Switzerland	295	76	13	384	34	1.6
Ten countries with known damage TOTAL	4 203	1 692	232	6 127	24	0.6
EUROPE				6 900	5	0.2

Source:

   Based on sources with varying methodologies. Data are not
strictly comparable between countries. For some countries the figures shown
are based on expert estimates or partial surveys (5).

*/   Asterisk indicates unofficial figure or secretariat estimate.

TABLE 2

Estimates of annual economic costs and benefits to forest industries of climatic change (95)

Cost/benefit in millions US dollars (1974) of
changes in mean global surface temperature

	- 1.0.C	+ 0.5.C
Forest production		
(Total) USA	-13 000	?
(Total) Canada	- 5 360	+ 3 620
(Total) USSR	-27 660	+12 600
Douglas Fir Production		
US Pacific north-west	- 9 500	?

Notes:

Negative sign denotes benefits; positive sign denotes costs.

Figures for -1.0.C will increase by 20 per cent for a 12.5 per cent decrease in precipitation, and vice versa.

Figures for + 0.5 per cent will increase by about 15 or 20 per cent for a 6.25 per cent increase in precipitation, but decrease by 80 or 90 per cent for a 6.25 per cent decrease in precipitation.

REFERENCES

1. Report of the third session of the Executive Body for the United Nations Economic Commission for Europe, Convention on Long-range Transboundary Air Pollution (ECE/EB.AIR/7) (1985).

2. Airborne Sulphur Pollution: Effects and Control, Air Pollution Studies No. 1 (ECE/EB.AIR/2) (United Nations publication Sales No. E.84.II.E.8, New York, 1984).

3. Air Pollution Across Boundaries, Air Pollution Studies No. 2 (ECE/EB.AIR/5) (United Nations publication Sales No. E.85.II.E.17, New York, 1985).

4. "International Co-operative Programme for Forest Damage Monitoring", Working Group on Effects (EB.AIR/WG.1/R.12) (United Nations, 1985) and papers of ECE/FAO workshop held in Freiburg (Federal Republic of Germany), October 1985.

5. "Consequences of forest damage attributed to airborne pollution", European Timber Trends and Prospects to the Year 2000 and Beyond (TIM/EFC/AC.3/R.6) (1985).

6. "Effects of sulphur compounds and other air pollutants on soil" in: Transboundary Air Pollution: Effects and Control, Air Pollution Studies No. 3 (ECE/EB.AIR/8) (United Nations publication, Sales No. E.86.II.E.23, pp. 43-63, New York, 1986).

7. "Effects of sulphur compounds and other air pollutants on groundwater", in: Transboundary Air Pollution: Effects and Control No. 3, pp. 64-77, op. cit. (2)

8. "Effects of air pollutants on agricultural crops", in: Transboundary Air Pollution: Effects and Control, pp. 27-42, op. cit. (2)

9. P.J.W. Saunders, The estimation of pollution damage, (Manchester University Press, 1976).

10. S.B. McLaughlin, "Effects of air pollution damage on forests: a critical review", Journal of the Air Pollution Control Association (35 (5), pp. 512-534, 1985).

11. B. Pruiz, E.B. Cowling and P.D. Manion, "Effects of air pollution on forests", Critical review and discussion papers, Journal of the Air Pollution Control Association, (35 (9), pp. 913-924, 1985).

12. Papers presented to the Symposium on the effects of airborne pollution on vegetation, (Warsaw, Poland, November 1979, United Nations Economic Commission for Europe, Geneva, 1980).

13. R.M. Adams, S.A. Hamilton and B.A. McCarl, The economic effects of ozone on agriculture, (US EPA-600/3-84-090, Oregon, 1984).

14. "Current status of the United Kingdom model for economic evaluation of acid deposition", Group of Experts on Cost and Benefit Analysis, (United Nations Economic Commission for Europe (EB.AIR/GE.2/R.11) (1985).

15. L.J. Eerden, A.E.G. Tonneyk and J.H.M. Wijnands, Damage to agricultural production in the Netherlands by air pollution, (Study commissioned by the Netherlands Ministry of Housing, Physical Planning and the Environment, 1985).

16. R.A. Wellburn, Oxides of nitrogen and their impact upon vegetation, the environment and ecosystems, World Health Organization Regional Office for Europe (WHO/EURO; ICP/CEH 002/m71/7, Copenhagen, Denmark, 1985).

17. S. Huttunen, Effects of SO_x on vegetation, WHO/EURO (ICP/CEH 002/m71/8, Copenhagen, Denmark, 1985).

18. Effects of ozone and photochemical oxidants on plants, WHO/EURO (ICP/CEH 002/m71/9, Copenhagen, Denmark, 1985).

19. Memoranda submitted by the Natural Environment Research Council and other bodies to the United Kingdom House of Commons, in: Evidence Submitted to and Report and Recommendations of the House of Commons Environment Committee Inquiry into Acid Rain, (HMSO, London, United Kingdom, 1984).

20. A.S. Lefohn and D.P. Ormrod, A review and assessment of the effects of pollutant mixtures on vegetation - research recommendations, (US EPA-600/3-84-037, Oregon, United States of America, 1984).

21. S. Schrader, H.R. Schinwald and D. Dujesiefken, Immissionen und Waldschäden Bibliographie - III. 1984 (in German), (Mitteilungen der Bundesforschungsanstalt für Forst- und Holzwirtschaft (140), p.183, Hamburg, Federal Republic of Germany, 1985).

22. S. Beilke and A.J. Elshout, eds, Acid deposition, Proceedings of CEC Workshop held in Berlin (West), September 1982 (CEC Brussels, Belgium, 1983).

23. D.G. Kinniburgh and W.M. Edmunds, The susceptibility of UK groundwaters to acid deposition (Report to DOE by NERC British Geological Survey, Wallingford, United Kingdom, 1984).

24. P.G. Whitehead and N. Christophersen, eds, SWAP Modelling Workshop Report, (NERC Institute of Hydrology, Wallingford, United Kingdom, 1985).

25. P.G. Whitehead, R. Neale and C. Neal, Predicting the effects of acid deposition on water quality (Second report, NERC Institute of Hydrology, Wallingford, United Kingdom, 1985).

26. K.M. Nyomárkay and others, Effects of SO_2 on plants, (English edition) (Gidrometeoizdat, p.136, Moscow, USSR, 1984).

27. Proceedings of World Climate Conference: A Conference of Experts on Climate and Mankind, (World Meteorological Organization No. 537, Geneva, Switzerland, 1979).

28. Proceedings of International Hydrological Programme Workshop - Hydrological and hydrochemical mechanisms and model approaches to the acidification of ecological systems (Uppsala, Sweden, September 1984).

29. G.A. Fraser and others, The potential impact of long-range transport of air pollutants on Canadian forests, Information report E-X-36 (Canadian Forestry Service, Alberta, Canada, 1985).

30. M.H. Unsworth and D.P. Ormrod, eds, Effects of Gaseous Air Pollutants in Agriculture and Horticulture (Butterworths, London, 1982).

31. M.J. Koziol and F.R. Whatley, eds, Effects of accumulation of air pollutants in forest ecosystems (Reidel, Dortrecht, 1983).

32. J.B. Mudd, and others, "Pollutants and plant cells: effects on membranes", in: Gaseous Air Pollutants and Plant Metabolism, eds. M.J. Koziol and F.R. Whatley (pp. 105-116, Butterworths, London, 1984).

33. D. King, "Modelling the effect of drought on crop sensitivity to ozone", in: 1983 NCLAN Annual Report (United States Environmental Protection Agency, Corvallis, 1984).

34. S.E. Lindberg and others, in: D.W. Johnson, ed., Walker Branch Watershed Synthesis (United States of America, 1985).

35. Report of the first session of the ECE Working Group on Nitrogen Oxides (EB.AIR/WG.3/2) (1985).

36. "Preliminary assessment of costs and benefits of reducing emissions from large firing installations in the Federal Republic of Germany", ECE Group of Experts on Cost and Benefit Analysis (EB.AIR/GE.2/R.13) (1985).

37. J. Bossavy, in: Air Pollution: Proceedings First European Congress on the influence of air pollution on plants and animals, Centre for Agric. Publ. and Doc., (Wageningen, Netherlands, 1969).

38. Various: primarily seeking to estimate total areas of western Europe exposed to ambient concentrations or deposition of sulphur, nitrogen and oxidant pollutants at levels suspected to inhibit growth and development of crops and other plants, or to contribute excess acidity to aquatic systems.

39. W.W. Heck and others, National Crop Loss Assessment Network (NCLAN) 1982, Annual Report. (United States Environmental Protection Agency, Corvallis, 1983).

40. D. Fowler and J.N. Cape, in: reference 30 op. cit. (1982).

41. D. Fowler and J.N. Cape, "On the episodic nature of wet deposited sulphate and acidity", Atmospheric Environment, 18, (pp. 1859-1866, 1984).

42. R. Crossley, NERC Institute of Terrestrial Ecology, (Edinburgh, United Kingdom, personal communication, 1985).

43. For instance: D.H. Peirson, and others, "Trace elements in the atmospheric environment", Nature, 241, (252-256, 1973) and subsequent reports by the same authors from AERE, Harwell and in the scientific literature.

44. D. Fowler, "Transfer to terrestrial surfaces", Philosophical Transactions of the Royal Society (London, 305, B. pp. 281-297, 1984).

45. D.W. Johnson, J. Turner, and J.M. Kelley, "The effects of acid rain on forest nutrient status", Water Resources Research, 18, (p. 449, 1982).

46. F.T. Last, D. Fowler and P.H. Freer-Smith, "Die Postulate von Koch und die Luftverschmutzung", (in German) Forstwissensch.Centralblatt, 103, (pp. 28-48, 1984).

47. B. Prinz and H. Stratmann, "Possibility of the analytical determination of dose-threshold value relationships", Staub-Reinhalt.Luft 30, (pp. 372-375, 1970, English translation).

48. R. Zahn, "The significance of continuous and intermittent sulphur dioxide action for plant reaction", Staub-Reinhalt.Luft, 23, (pp. 343-352, 1983, English translation).

49. J. Hajduk et al. "Einwirkungen von Industrieexhalationen auf die Struktur der Phytocoenosen", Gesellschaftsmorphologie, (pp. 340-349, 1970, in German).

50. W.B. Yapp, Production, pollution and protection. Wykeham Science Series (19), (London, 1972).

51. D. Mueller-Dombois, "Canopy dieback and successional processes", Pac. Sci., 37, (pp. 317-325 and pp. 483-486, 1983).

52. A. Baumgartner, "Climatic variability and forestry", in: reference 27 op. cit. (pp. 581-607, 1979).

53. R.L. Heath, "Initial events in injury to plants by air pollutants", Annual Review of Plant Physiology, 31, (pp. 395-431, 1980).

54. J.B. Mudd and others, "Pollutants and plant cells: effects on membranes". In: reference 32 op. cit. (pp. 105-116, 1984).

55. I. Ziegler, "The effects of SO_2 pollution on plant metabolism", Residue Reviews 56, (pp. 79-105, 1975).

56. J.S. Jacobson, "Ozone, and growth and productivity of agricultural crops". In: reference 30 op. cit. (pp. 293-304, 1982).

57. V.J. Black and M.H. Unsworth, "Effect of low concentrations of sulphur dioxide on net photosynthesis and dark respiration of Vicia faba", Journal of Experimental Botany, 30, (pp. 473-483, 1979).

58. A.R. Wellburn, "Effects of SO_2 and NO_2 on metabolic function", In: reference 30, op. cit. (pp. 169-187, 1982).

59. T.A. Mansfield and P.H. Freer-Smith, "Effects of urban air pollution on plant growth", Biological Review, 56 (pp. 343-368, 1981).

60. A.J. Zeevart, "Some effects of fumigating plants for short periods with NO_2", Environmental Pollution, 11 (pp. 97-108, 1976).

61. M.E. Whitmore, "Relationship between dose of SO_2 and NO_2 mixtures, and growth of Poa pratensis", New Phytologist, 99 (545-553, 1985). (Also, many others by Mansfield, Ashenden, Wellburn, et al.).

62. J.S. Jacobsen and J.J. Troiano, "Development of dose-response functions for effects of acidic precipitation on vegetation", Water Quality Bulletin, 8 (pp. 67-71 and p. 109, 1983).

63. A.S. Heagle, "Interactions between air pollutants and plant parasites", Annual Review Phytopathology, 11 (pp. 365-388, 1973).

64. R.M. Cox, "The sensitivity of plant reproduction to long-range transported air pollutants; in vitro sensitivity of pollen to acidity", New Phytologist, 95 (p. 269, 1983) (and others).

65. T. Keller and H. Beda, "Effects of SO_2 on the germination of conifer pollen", Environmental Pollution, 33 (pp. 237-242, 1977).

66. D. Drablos and A. Tollan, eds., Ecological impact of acid precipitation, Proceedings of an International Conference, (SNSF Project, Oslo, Norway, 1980).

67. D. Fowler, and others, "The influence of a polluted atmosphere on cutical degradation in Scots Pine (Pinus sylvestris)". In: ref. 66 op. cit. (p. 146, 1980).

68. T.M. Roberts, "Long-term effects of sulphur dioxide on crops: an analysis of dose-response relations", Philosophical Transactions of the Royal Society (London, 305 B, pp. 299-316, 1984). Also: Roberts et al, Environmental Pollution A, 39 (pp. 235-266, 1985).

69. D.W. Cowling, L.H.P. Jones and D.R. Lockyer, "Increased yield through correction of sulphur deficiency in ryegrass exposed to sulphur dioxide", Nature, 243 (pp. 479-480, 1973) (and others).

70. J.N.B. Bell, "Sulphur dioxide and the growth of grasses", In: reference 30 op. cit. (pp. 225-246, 1982).

71. P.C. Pande and T.A. Mansfield, "Responses of spring barley to SO_2 and NO_2 pollution", Environmental Pollution, A, 38 (pp. 281-291, 1985).

72. H.E. Heggestad and J.H. Bennett, "Photochemical oxidants potentiate yield losses in snap beans attributable to sulphur dioxide", Science, 213, (pp. 1008-1010, 1981).

73. A.H. Buckenham, M.A.J. Parry and C.P. Whittingham, "Effects of aerial pollutants on the growth and yield of spring barley", Annals of Applied Biology, 100 (pp. 179-187, 1982).

74. J.G. Horsfall and E.B. Cowling, eds., Plant Disease (Academic Press, New York, 1980).

75. R.I. Bruck, "Effect of simulated acid precipitation on the interactions of tree roots, soil-borne pathogens, ectomycorrhizal symbionts, and ecotmycorrhizae of loblolly pine", In: National Acid Precipitation Assessment Programme Rev. Doc. (Raleigh, North Carolina, United States of America, pp. 29-40, 1983).

76. G.R. Port and J.R. Thompson, "Outbreaks of insect herbivores on plants along motorways in the United Kingdom", Journal of Applied Ecology, 17, (pp. 649-656, 1980).

77. G.P. Dohmen, "Secondary effects of air pollution; enhanced aphid growth", Environmental Pollution, A, 39 (pp. 227-234, 1985).

78.   Z. Sierpinski, "Schädliche Insekten an jungen Kiefernbeständen in Rauchschadensgebieten in Oberschlesien", Archiv fir Forstwesen, 15 (pp. 1105-1114, 1966) (in German).

79. Cf. Fl ickiger, Hughes, Przybylski, Wentzel and Wiackowski.

80. The nitrogen cycle of the United Kingdom. A Study Group report, (Royal Society, London, United Kingdom, 1984).

81. S.I. Nilsson, H.G. Miller and J.D. Miller, "Forest growth as a possible cause of soil and water acidification: an examination of concepts", Oikos, 39 (pp. 40-49, 1982).

82. C.T. Driscoll and G.E. Likens, "Hydrogen ion budget of an aggrading forested ecosystem", Tellus, 34 (pp. 283-292, 1982).

83. D.W. Johnson, and others, "The effects of atmospheric deposition on potassium, calcium, and magnesium cycling in two deciduous forests", Canadian Journal of Forest Research cf. ref. 10, (1986).

84. B. Ulrich, R. Mayer and P.K. Khanna, "Chemical changes due to acid precipitation in a loess-derived soil in central Europe", Soil Science, 130 (pp. 193-199, 1980).

85. A. Ríhling, and others, "Fungi in metal contaminated soil near the Gusurn brass mill, Sweden", Ambio, 13, (pp. 34-36, 1984). Also papers by Bengtsson, Rundgren, Martin, Coughtrey, etc.

86. For example: D. von Eckstein, R.W. Aniol and J. Bauch, "Dendro- climatological investigations on fir dieback", European Journal of Forest Pathology, 13, (pp. 279-286, 1983), and A.H. Johnson, and others, "Recent changes in patterns of tree growth rates in the New Jersey pinelands: a possible effect of acid rain", Journal of Environmental Quality, 10, (pp. 427-439, 1981).

87.   Press release from World Climate Conference, 1985, World Meteorological Organization (Geneva, Switzerland, 1985).

88. D.M. Cushing, "Climatic variation and marine fisheries", In: ref. 27 op. cit., (pp. 608-627, 1979).

89. P. Ottestad, "Forecasting the annual field in sea fisheries", Nature, 185, (p. 183, 1960).

90. S. Zupanovitch, "Causes of fluctuation in sardine catches along the eastern coast of the Adriatic Sea", Anali Jadranskog Instituta, IV, (pp. 401-498, 1968).

91. J.D. McQuigg, "Climatic variability and agriculture in the temperate regions", In: ref. 27 op. cit., (pp. 406-425, 1979).

92. For example: J.R. McClenahen and L.S. Dochinger, "Tree ring response of White Oak to climate and air pollution near the Ohio River valley", Journal of Environmental Quality, 14, (pp. 274-280, 1985).

93. P.D. Manion, Tree disease concepts (Prentice Hall, Englewood Cliffs, New Jersey, United States of America, 1981).

94. European Communities Environment Protection and Climatology Research Programme: Concerted Action (COST) on "Atmospheric pollution effects in terrestrial and aquatic ecosystems". Various publications and reports. (CEC, Brussels, Belgium, 1984/1985).

95. R.C. d'Arge, "Climate and economic activity", In: ref. 27 op. cit., (pp. 652-681, 1979).

96. J.L. Schnoor, L. Sigg and W. Stumm, "Acid precipitation and its influence on Swiss lakes", EAWAG News, 14/15, (pp. 6-12, 1983).

97. C.O. Tamm, L. Hallbäcken, "Changes in Soil Acidity from 1927 to 1982-84 in a Forest Area of south-west Sweden", submitted for publication 1986.

98. T.C. Hutchinson, M. Dixon and M. Scott, "The effect of simulated acid rain on feather mosses and lichens of the boreal forest", Water, Air and Soil Pollution, (1986).

99. J. Nilsson, ed. "Critical loads for Nitrogen and Sulphur", Nordisk Ministerräd, (Miljo raport No.11, 1986).

100. G. Agren, "Model analysis of some consequences of acid precipitation on forest growth". In: Ecological effects of acid deposition, (National Swedish Environmental Protection Board Report 1636: 233-244, 1983).

Chapter 2

EFFECTS OF AIR POLLUTANTS ON AQUATIC ECOSYSTEMS

113. In humid temperate and boreal areas, the paludification and podzolization of catchments may cause slow natural acidification of sensitive lakes and rivers, a process accelerated and widened by human activities. Anthropogenic emissions of atmospheric pollutants add significant quantities of acid compounds to the environment. Emissions of sulphur dioxide (SO_2) increased between 1950 and 1970 by a factor of two in most of Europe and parts of North America, and then generally stabilized. In the eastern United States this increase continued into the early 1980s while, in contrast, emissions of SO_2 declined markedly in most parts of western Europe.

114. Data on nitrogen oxides (NO_x) are less comprehensive but up to 1970 the trends generally follow those of SO_2. From then on, emissions further increased with the growth in the number of motor vehicles into the 1980s. A similar situation also prevails for hydrocarbons (an important precursor in ozone production).

115. In some regions of Europe and North America in terms of wet deposition, sulphuric acid and nitric acid contribute respectively 60 to 70 per cent and 30 to 40 per cent approximately to the total acid in deposition; the relative contribution of nitric acid has increased during the past 30 years. In some countries there has been a doubling of emissions in this time period. The acidifying potential of wet deposits is also influenced by the concentrations of ammonia and base cations.

116. The rate of wet deposition of sulphur ranges from less than 3 kg S ha^{-1}yr^{-1} in northern Scandinavian countries to more than 30 kg S ha^{-1}yr^{-1} in rural areas in central Europe. The corresponding figures for inorganic-nitrogen are in the range of 1 to 3 kg N ha^{-1}yr^{-1} in northern Scandinavian countries compared to measured deposition values of more than 20 kg N ha^{-1}yr^{-1} over large areas of central Europe and North America.

117. Ammonia can also greatly influence the acidifying potential of rain; although it neutralizes acids in rainwater, ammonia is eventually converted into nitrate by nitrifying bacteria in the soil. This process produces two H^+ ions for each NH_4^+ ion nitrified.

I. PROCESSES FOR AQUATIC ACIDIFICATION

A. GEOCHEMICAL PROCESSES

118. H^+ is a highly reactive agent and so its geochemical behaviour differs markedly from that of a weakly interacting pollutant such as nitrate. H^+ ions are unlikely to pass through soils or rocks without reaction; consequently they will tend to be absorbed. The soil provides the principal short-term sink for much of the externally and internally produced acidity.

119. Krug and Frink (1) observed that most acid precipitation research did not pay enough attention to interactions of acid rain, acid soil and vegetation. They also hypothesized that an increase in SO_4^{2-} flux would be balanced by a decrease in organic anion flux, resulting in little or no change in pH. However, although organic anions may be important, there seems little experimental evidence to support this contention. This explanation is obviously insufficient to explain regional acidification (Seip, (2); Henriksen (3)). The subject has led to active debate which is not yet resolved. (Lazerte and Dillon (4)).

120. In the studies carried out under the SNSF project (the Norwegian Interdisciplinary Research Programme), little evidence was found of an important effect on freshwater acidity from

land-use changes (5). This was not the case with some other studies where land use and management practices were found to be influential factors. There is evidence that afforestation and deforestation (6; 7; 14) have significant effect on the chemistry of run-off water. In Sweden, forest fertilization has also increased the nitrate ratio in run-off waters (8). Nitrate leaching is increasing also in non-fertilized forests, reflecting the effects of increased atmospheric deposition of nitrogen.

121. The possible importance of forest soil processes in defining temporary surface-water pH depression was studied by Lefohn and Klock (9). An assessment of natural biogeochemical processes indicated that H^+ ions were produced in temperate forest ecosystems by nitrification, oxidation of sulphates and by dissociation of organic weak acids. The geological origin of sulphates may be significant in some locations. In areas receiving substantial sea salt inputs, temporary acid releases were observed (11, 12).

122. Statements cited from recent literature on soil-water interactions point to the importance of further research and to the results expected from the International Co-operative Programme.

B. BIOLOGICAL PROCESSES

123. Biological processes in soil lead to the production of natural acids such as HCO_3^- and organic acids. CO_2 is formed by decomposing plant materials. At the same time basic ions Ca, Mg, K and Na are released. Input-output budgets and estimations of natural proton generating processes show generally that natural acids play a minor role in the relation between acid deposition and acid surface waters (13). However, it should be recognized that the effect of strong acids could be significantly different from that of weak acids, commonly formed in soils. Ion balances in the Solling Forest of the Federal Republic of Germany show that less than 30 per cent of the total acid generation derives from natural sources (14), but it could be higher in less polluted areas depending on site conditions.

II. EFFECTS ON WATER CHEMISTRY

A. RELATIONSHIP BETWEEN ACIDIFYING DEPOSITION AND WATER QUALITY

127. Freshwater acidity is a result of complex interactions between wet and dry deposition, soil

124. A significant sulphur input to some surface waters may occur because of input from water-logged soils where biogenic-reduced sulphur compounds are concentrated. This reduced sulphur may then be oxidized to sulphate in sediments of lakes and rivers, or in soils when the water table drops.

125. Bacterial processes which consume hydrogen ions while reducing nitrate and sulphate were studied in lakes in southern Norway, the Adirondack Mountains (United States of America), southern Ontario (the Haliburton region of Canada) and in north-western Ontario (Canada) (experimental lakes area - ELA). Measurement of rates of nitrate and sulphate reduction and profiles of sediment porewater chemistry clearly demonstrated that nitric and sulphuric acids were actively consumed in the epilimnetic sediments of all of these lakes. Acidification stimulated nitrate and sulphate reduction because nitrate and sulphate concentrations had been elevated by acid inputs. These processes play only a special role in lakes with anaerobic hypolimnia.

126. The duration of acidification does not appear to affect the activity of the nitrate- and sulphate-reducing bacteria. Sediments from recently acidified ELA lakes (10 years) reduce nitrate and sulphate at about the same rates as sediments from atmospherically acidified lakes where the pH had been reduced for many decades. In the Adirondack Mountains, where the nitrate concentrations were elevated, alkalinity production by denitrification exceeded the production of alkalinity by sulphate reduction processes by about one third. In the Haliburton Lakes where nitrate concentrations were very low, sulphate reduction was primarily responsible for alkalinity production. In the Norwegian lakes studied, which had highly organic sediments, this inefficiency was probably due to differences in the end-products of sulphate reduction and to conditions within the sediments which promoted re-oxidation of reduced iron sulphides. The differences between these two sediment types may have an important bearing on the capacity of lakes to neutralize acid inputs, and to recover once acid deposition is reduced (15).

and rock conditions, land use and hydrology of catchment. Hydrological conditions and biogeochemical processes affect the run-off composition and subsequently the quality of freshwater.

128. Sulphur from deposition is transferred to aquatic systems through soils which may retain and later release the sulphur. At present these processes are not well understood in a quantitative way. In some cases there is evidence of an overall balance of input versus output on an annual basis (Birkenes model (16)). In the shorter term, discrepancies have been observed.

129. Wright (17) reported input-output budgets at Langtjern, a small acidified (pH 4.6-4.8) lake in southern Norway. The seven-year budgets for major ions at two subcatchments indicated that outputs greatly exceeded inputs for Ca^{2+}, Mg^{2+} and Al^+. Outputs were much less than inputs for H^+, NH_4^+ and NO_3^- while there was a rough input-output balance for Na^+, K^+, SO_4^{2-} and Cl. The catchment might retain about 20 per cent of the incoming sulphate (dry/or bulk deposition included). For the lake itself, H^+ input exceeded output; the same may be the case for NH_4^+ and NO_3^-.

130. Sensitivity of a lake or river depends on the acidity and amount of deposition plus other factors such as soil characteristics of the drainage catchment, the canopy effects of the ground cover and the composition of the bedrock. The capacity to buffer incoming acids is reflected in the alkalinity production from the catchment. Soft-water lakes, which are most sensitive to additions of acid substances, are usually found in areas with igneous bedrock which contributes few soluble solids to surface waters. Hard waters contain large concentrations of alkaline earths, chiefly bicarbonates of calcium and magnesium, derived from limestones and calcareous sandstones in the drainage catchment. Near coastal areas, marine salts may also be important in determining the chemical composition of a stream, river or lake.

131. For lakes, Henriksen (3) uses a titration curve for a bicarbonate solution to infer an empirical model of acidification. This model describes the pH of freshwaters on the basis of a site-specific intrinsic parameter related to sensitivity (Ca^{2+} and Mg^{2+} concentrations in water) and an external parameter related to loading of strong acid - SO_4^{2-} concentration in lake water. The model is based on the theoretical acid-base titration, and it works for several independent sets of data from Norway and for some other acidified areas (18). This model has been further developed (Wright, Henriksen) to take account of increased weathering of Ca and

Mg in acid conditions. The diversity of lake responses (with similar deposition loading) may be explained by differences in weathering yield of bases.

132. In the case of nitrogen, adsorption does not appear to be significant in soils, but there will be biological retention and release. A distinction needs to be made in the acidifying role of ammonia with its salts, and inorganic nitrates. Biological activity can affect either and convert between them. Fixation of nitrogen gas from the air can act as another source of nitrogen for soil and aquatic systems, while denitrification processes can provide a sink.

133. Nitrate concentrations in water from forested areas are usually low; during periods of rapid snowmelt or run-off, and after land-use disturbances, nitrate concentrations may rise. In these conditions, it serves as a mobile anion carrying cations, e.g. aluminium, through to surface waters. Several studies have been undertaken in Scandinavian countries and elsewhere, on assessment of effects of nitrogen deposition on the acidification of terrestrial and aquatic ecosystems (19).

134. In some locations, ammonium deposition is a significant component of the N input to surface water, for example 25 per cent in Harp Lake, Ontario (Canada) (20), in the Netherlands (over 50 per cent) and in Denmark (see Figure 1).

135. It has been shown in forested catchments in southern Sweden that nitrate input exceeds vegetative uptake while high nitrogen losses (6.2-10.9 kg ha^{-1}) were registered from spruce-dominated areas. Most of the nitrogen was in the form of NO_3^- and the highest NO_3-N content in stream water observed was 8.1 mg.l^{-1} (19). Aluminium levels were found to be high (up to 4 mg.l^{-1}) for a short period. These data indicate a direct association between Al^+ and NO_3 leaching in certain acid soils. This is also evident from lysimeter studies.

136. The amounts and forms of aluminium leached from soils and reaching surface waters will be determined by the chemical and physical conditions in soils, sediments and surface water on the principle that there is a connection between pH valve and Al concentration. The chemical equilibria determining aluminium speciation is the subject of current research activity but there is difficulty in relating theory (e.g. equilibrium with gibbsite) with field measurements. Nilsson (21) has suggested that inorganic aluminium in soil solution is regulated by formation of basic aluminium sulphates.

Figure 1

**Monthly changes in ammonium and nitrate concentration in water
at EMEP station in Keldsnor (Denmark)**

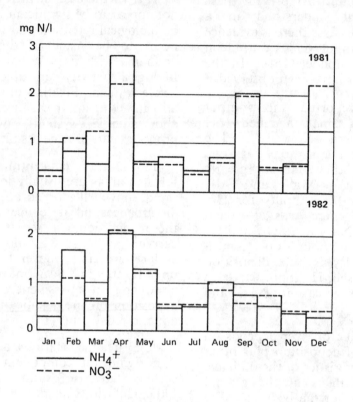

Figure 2

A nomograph to predict pH of lakes (Henriksen 1980)

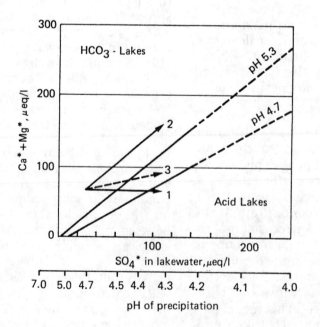

137. Driscoll et al (22) have studied aluminium chemistry in several acid lakes and streams in the Adirondack Mountains of New York State (United States of America). Aluminium solubility appeared most often to be regulated by an aluminium trioxide phase slightly more soluble than gibbsite. Fifteen per cent of all samples was under-saturated with respect to the solubility of synthetic gibbsite. These samples were often associated with rainfall or snowmelt events, when equilibrium conditions are not present.

138. Other soil-derived metals, e.g. Fe, Mn, Cd, Zn, etc. are also mobilized by acid soil conditions. On reaching surface waters they may have biological implications. Heavy metals from polluted deposition are for the most part retained in soils or lake sediments. The presence of mercury in lake sediments, its release to the water and transfer to biota is of continuing interest in acidified areas (23).

B. SHORT-TERM EPISODES IN WATER CHEMISTRY

139. Normally, surface waters are less acid than precipitation, but in areas of forest, unmanaged moorland or where peat deposits are common, the run-off may be more acid than the rain. Following snowmelt or heavy rain, run-off may be more acid than precipitation (5). Lefohn and Klock (9) have suggested that the observed increase in run-off acidity during storm events and snowmelt may be more a function of key climatic and biogeochemical processes than of precipitation chemistry.

140. Records of short-term fluctuations in the chemistry of acid waters are published for a number of locations (e.g. Norway (24)). In this case a contemporaneous fish kill was recorded; the pH was 5.1 and labile aluminium 120 ug l^{-1}. Entrainment of sea salts into deposition is also reported to result in acid increases in surface waters (United Kingdom, (25)).

141. It is important to define the amplitude, time scale and frequency of acid events, so as to understand their biological effects. The recent development of automatic field monitoring techniques will provide extensive and more detailed records of acid fluctuations in upland streams.

C. GEOGRAPHICAL EXTENT OF LAKES AND STREAMS AFFECTED BY ACIDIFICATION

142. Declines in pH or alkalinity have occurred in some surface waters over broadly distributed regions of Europe and North America. The changes in aquatic chemistry have in some cases led to reductions in, or the disappearance of, fish populations. Historical field evidence from Scandinavian countries, Canada and the United States (Adirondacks) as well as confirmation from laboratory and field studies indicate the mechanisms by which changes in aquatic chemistry can harm both adult fish and fish reproduction (20). In some areas sufficient data do exist to reach consensus on how many lakes and streams have been significantly altered or how many will change, for better or worse for fish, at current deposition levels.

143. Wright (18) reviewed freshwater acidification in Europe, where this phenomenon is not limited to Scandinavia. Surface waters in Finland, the United Kingdom and central Europe are now reported as affected (26, 27). In the Italian Alps, Mosello and Tartari (28) have reported pH and alkalinity for 320 alpine lakes. In most of the alpine region acidification is a less serious problem than in northern Europe. Some 95 per cent of the lakes had pH above 6 but 65 per cent had alkalinity below 100μ eq l^{-1} making them sensitive during snowmelt.

144. Measurements of ion transport through a forested catchment in the Hartz Mountains have been used in a hydrological model (Hauhs (7)).

145. In central Europe, poorly buffered lakes are found in areas of granitic gneissic bedrock, and thin sandy soils. Diatom and cladoceran remnants have indicated a pH change in this century (29). Other issues and surveys are included in work of Zöttl (30), Schoen et al (31) and several other publications (10, 25, 32). The current chemical and biological status of major Yugoslav surface freshwaters with respect to acidification has been assessed (33). Chemical results showed relatively high acid neutralizing capacity but sulphur loads throughout the country remain a concern. In Hungary, acidification effects on surface waters have been studied in the Csórrét catchment (34), in general, the natural buffering capacity of surface waters in the country is high owing to high HCO_3^- concentrations. Alkalinity is usually higher than 2.5 meq.l^{-1}. Nevertheless, in some reservoirs situated in the Matra Mountain region, the water has low alkalinity and pH values. Many rivers, streams and lakes in Poland contain large amounts of Ca^{2+} and Mg^{2+} so that they are less sensitive to acidification, but some surface waters are showing the effects of acidification (35).

146. In the United Kingdom, some acid lakes are found in south-west Scotland and west Wales. There are acid upland streams in west-central and south-west Wales, southern Pennines, Cumbria and central Scotland (25). A trend to increasing acidity is limited, however, to records of sediment cores from five lakes in south-west Scotland (36, 37) and of changed water quality in a few upland streams in Wales over the past 20 years. In contrast, upland towns in the Lake District show no significant changes over the past 40 years.

147. In Belgium and the Netherlands, acid moorland pools are found lying in sandy soils. These waters are usually of higher conductivity and nutrient concentration than water in granitic areas (26, 38, 39).

148. In Finland, a survey of 9,000 lakes (16 per cent of the total number of lakes > 1 ha) show that in southern Finland there are about 450 acid clear water lakes with pH value < 5.0 but of these only 160 are not humic. The acid clearwater lakes are seepage lakes or supplied by headwaters with thin soils catchments situated in areas of granitic bedrock. Paleolimnological studies have shown that many of these lakes have acidified during the last few decades (111).

149. In Norway and Sweden, monitoring programmes and regional surveys provide additional results to depict the well-described situation (41, 42).

150. An overview of freshwater acidification was given for Canada by Dillon (43). He concluded that there was considerable evidence of long-term increases in the SO_4^{2-} concentrations of lakes and rivers, some evidence of long-term decreases in pH and acid-neutralizing capacity, and some indirect evidence that cation concentrations had increased.

151. In the United States, the Environmental Protection Agency (EPA) is carrying out surveys in areas sensitive to acidification. A report was published in 1976 (112). Data from approximately 2,500 streams and lakes have been mapped with apparent spatial correlations between these data and macro-catchment characteristics, especially land-use (44). Chemistry and fisheries in Adirondack lakes have recently been reassessed and reported on (45). Numerous reports have recently appeared concerning acidification of lakes in Florida (46), in the West and in the Rockies.

D. TRENDS IN OBSERVED CHANGES

152. Historical trends in water quality are hard to establish quantitatively. Reliable chemical data are available only for the last 10 or 15 years at best; however, lake sediment cores and studies of diatom and other remnants in sediment cores can be used to reconstruct a possible past water quality.

153. Battarbee (36, 37) has studied cores from seven lakes in south-west Scotland, and has inferred a pH decline from 6 to about 4.5 within the past 130 years, some quite recently. Davies et al (47) have studied sediment cores from nine lakes in southern Norway and from northern New England (United States of America), inferring decreases of 0.6 to 0.8 pH units since the beginning of the century for the three Norwegian lakes with present pH values of less than 5.0. The six New England lakes decreased by 0.2 to 0.4 units. Tolonen and Jaakkola (48) studied sediment cores from four small lakes in southern Finland finding evidence for acidification in recent years. In one lake pH decreased by 1.1 to 1.5 units, starting around 1962. Steinberg et al (49) and Arzet et al (50) have studied diatom remains in dated sediment cores from six lakes draining granite and gneiss as well as in sandy lowland regions of the Federal Republic of Germany. In the Bavarian Forest, pH decreases of 0.6 to 0.8 units have been inferred during the last 30 to 80 years. Other biological remnants support the diatom influences (Kraus-Dellin and Steinberg (51)). Charles (52) has studied a sediment core from the Adirondack Mountains of New York State (United States of America) showing an inferred pH of about 5.7 up to 1950 then a decline subsequently to 4.7. Studies of sulphur in the sediments of a pond in the Adirondacks also indicated that significant increases in sulphur deposition did not occur until about 1930 but that loadings there had increased 2- to 3-fold during this century.

154. Shorter-term trends (10 to 15 years) are being studied. Smith and Alexander (53) looked at 10 to 15 years of stream chemistry data at United States Geological Survey Hydrological stations. At most south-eastern and western sites, a small increase was found in sulphate concentrations in streamwater. The regional pattern of stream sulphate trends is similar to that reported for SO_2 emissions during the same period. Trends in alkalinity display a geographic pattern that is the approximate inverse of that of sulphate trends.

155. Alasaarela and Heinonen (54) compared alkalinity of Finnish lake waters for the periods

1912-1930, 1962-1972 and 1972-1982. A general decreasing trend of alkalinity was common to all rivers.

156. Watt et al (55) found that pH in rivers in geologically sensitive areas of Nova Scotia (Canada) had decreased significantly over a 27-year period. Concomitant with this decrease were declines in HCO_3^- concentrations and increases in Al^{3+} and SO_4^{2-}. The authors attributed an average 73 per cent of the increase in acid to increased SO_4^{2-}. Clair and Witfield (56) studied 11 rivers in Atlantic Canada. Of eight sensitive rivers, four showed decreasing pH of which two showed increasing sulphate.

III. EFFECTS OF ACIDIFICATION ON BIOTA

159. Most freshwater plants and animals have a specific pH-tolerance range and disappear from a system if pH exceeds this range. In addition to such direct effects, organisms can disappear following the loss of their plant food or prey species and through increased competition. Acidification may decrease phosphorus and inorganic carbon concentrations, and depress nutrient cycling rates. These changes tend to decrease phytoplankton biomass and productivity. Acidification also may increase water clarity, allowing light to penetrate into deeper waters and tending to increase primary production. Evidence for both types of changes has been obtained in field studies (20). In general, however, acidified waters have reduced diversity of taxa.

A. MICRO-ORGANISMS

160. Little information is available on the role of micro-organisms in ecosystems affected by acid deposition. Sulphur bacteria have been shown in certain circumstances to contribute significantly to the acid neutralizing capacity of surface waters via sulphate reduction. Studies have indicated that the rate of microbial decomposition of organic matter is reduced in acid lakes and this can affect nutrient cycling and production at other trophic levels (61, 62). Decomposition of leaf litter exposed in situ was markedly increased after liming. In the acidified Lake Gardsjön (Sweden) mean bacterial cell size was comparatively large while bacterial biomass and phytoplankton biomass were comparable. SO_4^{2-} reduction occurred generally while NO_3^- reduction was negligible.

157. The decreasing trend in pH in many rivers in southern Norway found up to 1979 seems to have ceased. There is no clear trend in later years (57, 58). Small catchments' studies show consistent changes, but indicate some recent improvement in water quality (41). In Sweden some lakes have shown a decline in sulphate concentration but not always a comcomitant change in pH (Hultberg (59)).

158. Morgan (60) re-evaluated data for pineland acid streams in New Jersey (United States of America) and showed that the earlier conclusion of acidification in two streams was not verified.

B. PLANTS

161. The impact of acidification on phytoplankton and zooplankton communities has been reviewed by Geelen and Leuven (63) for Europe. Their review concluded that in acidifying systems there was a considerable decrease in diversity of both phytoplankton and zooplankton and a few species become dominant, but productivity is not decreased.

162. Broberg and Persson (64) studied phosphorus productivity regulatory function in acidified and limed lakes. Phytoplankton were regulated by phosphorus but did not reach the chlorophyll/phosphorus ratio observed in circumneutral lakes. Phosphate activity was high in the acid lakes. Substrates tended to react with Al^{3+} thereby causing excess enzyme production by algae and competition between phosphates and Al^{3+} for substrate.

163. After liming a lake, the species of epiphytic algae present changed and biomass decreased (61). Benthic algal mats covered large areas of the lake bed but two years after liming they had disappeared and **Lobelia** plants were recovering.

164. The extensive growth of filamentous algae **Mougeottia spp** is now considered to be a frequent phenomenon in acidified lakes, but the mechanisms involved are scarcely understood (65).

165. In acidified lakes of the Bavarian Forest, submerged macrophytes have been studied by Melzer et al. (66, 67, 68). They have shown that in all lakes **Juncus bulbosis f. fluitans** is the most

abundant spermatophyte and that several **Sphagnum spp** had invaded the benthos.

166. Roelofs (69) has described the changes in Western Europe that have taken place in the plant communities of heathlands and peatlands over the past decades and attributed to atmospheric sulphur and nitrogen deposition. The numbers of moorland pools and small lakes in which **Littorella** species occur has declined drastically.

C. MACRO-INVERTEBRATES

167. There is a good deal of work currently being carried out in different countries on the effects of acidification on invertebrates. Results substantiate earlier findings that as pH falls some groups of organisms become depleted or absent. **Dugesia gonocephala**, a typical planarian of mountain streams, do not occur in areas having pH 6.0 (70, 71). When exposed to pH 5.0, animals die within a few days. At very low pH the epidemies bursts and the cytoplasm is dissolved. In the Federal Republic of Germany, Norway and Sweden distribution of the freshwater shrimp **Gammarus** is a good indicator of headwater acidification (72, 73, 74). Ephemerotropa are missing in streams with pH value < 5.0.

D. AMPHIBIANS

168. Although little information is available, anurans and salamanders may be significantly affected by acidification. Reproduction of certain amphibians often takes place in temporary pools of melt water. In addition, the young of several species congregate in warm shore pools during spring. A combination of little buffering capacity and episodic inputs of acidic substances will result in elevated acidity at such localities in excess of the levels tolerated by several species of amphibians. Studies (75, 76, 77) have shown that amphibian populations have been lost, diminished in abundance, or restricted in distribution depending on the acidity of their habitat. Reproductive impairment in amphibians may be one of the primary effects of increased acidity.

E. FISH

169. The effects of acidification on fish populations have been investigated in some geographic areas. In Norway and Sweden, where lake surveys have been quite extensive, thousands of fishless lakes have been identified (78, 79). The number of fish species has been positively correlated with pH for the La Cloche Mountain Lakes, Ontario (Canada) (80). Regional surveys in Norway indicate that there is an increase in the area where damage to fish is observed (81).

170. A frequent observation in fish populations at low pH is the failure to recruit young into the population. Stress due to acidification is commonly associated with a drastic reduction in population size and biomass, often leaving a small number of large fish prior to extinction. The exact mechanism responsible for recruitment failure is not known; it may involve the impairment of ionic regulation or failure of larval emergence.

171. Jernelöv (82) studied the effects of acid rain on mercury levels in fish. The statistical correlation between low pH in lakes and methyl mercury levels in fish has been well established from investigation in Sweden, Finland, Canada and the United States. One of the mechanisms that can contribute to this phenomenon relates to the instability of dimethyl mercury in contact with acidity in the atmosphere and the resulting formation of water-soluble monomethyl mercury.

172. Effects of lime application on the aquatic biota in acidified lakes and streams in Sweden were studied by Eriksson et al. (83). A few years after liming, species composition and diversity of phytoplankton, zooplankton and benthic insects were almost identical to those found in oligothropic and non-acid lakes. Reproduction of remaining species of fish was successful as pH increased. The same was usually true for restocking of depleted fish stocks.

173. In Norway significant decline in fish stocks had occurred along the southern coast (Vest-Agder and Aust-Agder counties) from 1974/78 to 1983. By 1983 some 71 per cent of brown trout populations and 43 per cent of perch populations in the area had been lost. Episodic fish kills have been reported in the River Vikedalselva in south-western Norway, and in the Gaula River, where diversity of invertebrate fauna is also reduced (41).

174. In conclusion, there appears to be a need for quantitative regional estimates of current and future losses or recovery of fisheries resources resulting from surface water acidification and acid deposition. These are essential in order to evaluate the potential benefits of emission control. Laboratory and field experiments have confirmed that chemical conditions associated

with acidification (low pH and elevated aluminium levels in waters with low calcium concentrations) are toxic to fish. Determining actual numbers of fish populations lost to date or predicting future changes in the fisheries resource is much more difficult. Compilation and analysis of existing data bases for lakes most likely to have been affected by acid deposition in many cases provide semi-quantitative assessments of the extent and magnitude of impacts on fisheries.

F. BIRDS AND MAMMALS

175. Birds and mammals most probably are affected indirectly by acidification and acid deposition due to changes in their habitat caused by acid conditions and changes affecting the availability and quality of their food. Reduction or disappearance of organisms in the food chain could affect the food availability for many species such as waterfowl, herons, grebes and semi-aquatic carnivorous mammals. Long-term

effects may be noted, ranging from declining breeding densities to local extinction.

176. Eriksson studied lake acidification and bird populations in Sweden (84). Low production of young osprey, a fish eater, has been recorded in breeding areas with many acid lakes. In Wales (UK) Ormerod (85) has related loss in dipper populations (an insect feeder) to increased acidification of streams and afforestation.

177. Trophic relationships were studied in order to identify the mechanisms by which acid precipitation affects waterfowl productivity in aquatic ecosystems at different stages of acidification in Canada. The decline in the abundance and occurrence of fish in some areas had reduced the capacity of the habitat to support broods of fish-eating birds, such as common mergansers and common loons. However, insectivorous waterfowl such as common goldeneyes prefer fishless lakes and may derive some short-term benefits from reduced competition with fish for their common insect prey (86).

IV. MODELLING OF ACIDIFICATION PROCESSES AND POSSIBLE CONSEQUENCES

178. Models, empirical or mathematical, have been developed over recent years both to improve understanding of acidification processes and to predict the consequences of changed emissions.

179. Henriksen (87) presented a model for predicting changes in water chemistry (see Figure 2) and fishery status following changes in sulphur deposition (88,24). A 30 per cent reduction in sulphur concentrations of lake waters is expected to improve possibly about 22 per cent of acidified lakes in southern Norway which should be able to support fish. Some assumptions in these calculations are difficult to evaluate (89).

180. Kramer and Tessier (90) have made a critical review of some models of lake acidification. They have suggested some improvements regarding assumptions in these models. For instance, inclusion of Na^+ and K^+ in addition to Ca^{2+} and Mg^{2+} (89). Dickson (91) has discussed the relative importance of sulphuric and nitric acid for acidification of Scandinavian freshwater systems. Sulphate predominates in lake waters on an annual basis, but during spring flow, nitrate may have a great influence on acidification and aluminium mobilization. Modelling has been particularly important in the ILWAS (Integrated Lake Watershed Acidification Study) project. The comprehensive

ILWAS model has been described by Goldstein et al.(92).

181. Several recent models have important common features with respect to soil chemical processes. They more or less explicitly include the mobile anion concept, along with cation exchange, weathering, and dissolution equilibria for controlling aluminium concentrations.

182. Reuss and Johnson (11) investigated the relationships between soil solution alkalinity and the soil CO_2 partial pressure, $(CO_2)g$, at different strong acid anion levels. A major finding was that for a given $(CO_2)g$, an increase in strong acid anions can result in a switch from positive to negative alkalinity in the soil solution. At elevated $(CO_2)g$ levels in the soil, the pH of the soil solution will be only moderately affected, whereas the pH of the leachate (after CO_2 equilibration) will be strongly affected. This mechanism could be of particular importance in explaining observed reductions in freshwater pH.

183. Cosby and co-workers have used an extended version of the Reuss-Johnson model as a basis for MAGIC. One objective was to determine whether a lump sum representation of chemical reactions in soils would yield a satisfactory description of gross chemical

behaviour in catchments for use in long-term predictions (12, 93). Model results have been compared to historical pH trends inferred from diatom studies of lake sediments. Wright et al. (94) used the MAGIC model for the interpretation of paleolimnological reconstructions of soil and water acidification. In sensitive areas receiving acidic deposition, paleolimnological data indicate changes in lake pH over one to three decades during the past century. Estimates of emissions and deposition of SO_x and NO_x over this same period suggest that deposition increased slower than did changes in lake pH. Chemical and biological processes in the terrestrial catchment and lake ecosystem moderate the response of surface water pH to deposition of acidifying compounds. The results indicated that the processes linked in MAGIC can account for temporal trends in pH such as those obtained from paleolimnological data.

184. Within the Norwegian project RAIN, a large-scale manipulation project, data are being collected on the effect on water chemistry of changing acid deposition in selected catchments (95). These data will be utilized in further modelling in future.

185. The "Birkenes model" has been developed further and applied to a stream in Ontario (96). Chemical data for a period of five years were simulated quite well. The model has also been used tentatively for predictions of response on water acidity to changes in sulphur deposition. By doubling or halving the sulphate concentrations in streamwater compared to present-day values, and assuming a constant time potential in the soil, peak acidity during snowmelt changed with up to 0.9 pH units; shifts of 0.4 to 0.5 units were predicted in many cases. Modified Birkenes models have also been used by Lam (97) and by Whitehead et al. (98, 99). The model has also been applied to the snowmelt process (16). In general the MAGIC and Birkenes model chemistry is very similar although modifications would be required in order to consider special factors such as sea salts, differing

aluminium speciation, enhanced biological activity and the effects of changing soil characteristics. A major difficulty in the development of such models is the lack of knowledge of weathering rates. To obtain better values is an important goal of many programmes.

186. A rather different approach has been taken by Schnoor and co-workers in developing the "Trickle Down" model. They base their models on mass balance for alkalinity. The rate of supply of alkalinity to water within a catchment is kinetically controlled. The steady-state version is described by Schnoor and Stumm (100) while a "time-variable" model for seepage lakes is described by Lin and Schnoor (101).

187. Alcamo et al. (102) have developed the RAINS (Regional Acidification Information and Simulation) model. The principal purpose of the model is to provide a tool to assist decision-makers in their evaluation of strategies to control acidification of Europe's environment. Parts of this model system are also described by Kauppi et al. (103), Posch et al. (104), and by Kamari et al. (105). The report by Kamari et al. describes a model for predicting long-term lake-water acidification on a large regional scale. Some important assumptions are similar to those in the Birkenes model. Monte Carlo techniques are used to determine ranges and combinations of input values that produce acceptable results. A simple semi-empirical model for soil acidification and ranking of sensitivity of soils to acidic deposition has been developed by Bloom and Grigal (106). A dynamic model for the long-term effects of acid deposition is being developed at the International Institute of Applied Systems Analysis (IIASA) in Laxenburg (Austria) (107, 108, 109).

188. There is increased interest in identifying critical loadings of acidifying pollutants, in order to protect the sensitive surface water. The further collection of data and development of models will help to resolve present differences of views.

V. CONCLUSIONS AND RECOMMENDATIONS

189. A relationship between acid deposition and water and soil acidification has been established in many areas. There is also evidence that this acidification is directly related to negative effects on aquatic ecosystems in a number of areas of Europe and North America.

190. The extent of acidification of surface waters varies from region to region. Acidification is limited at present to water areas in non-calcareous rock and sandy soil. These conditions - decreases in pH, or alkalinity - are found over widespread areas in Europe and North America.

191. Soil buffering-capacity drops with increasing acidification of the soil. This results in increased acidification of groundwaters, especially where aquifers are shallow. Acidification of an aquifer results particularly in the release of aluminium. An early indicator of potential groundwater acidification is acidification of surface waters.

192. Biotic and chemical changes have been observed in affected areas as a result of water acidification. This can lead to fish loss and complete desolation of waters.

193. The chemistry of aluminium plays a major role in the acidification of surface waters, groundwaters and biota. However, in order to be able to determine the degree to which the rate of acidification affects the leaching of aluminium, the factors controlling aluminium solubility and speciation need to be better understood.

194. Although information on the impact of nitrogen deposition on aquatic ecosystems is scarce, it is clear that nitrogen, which is normally a nutrient, may create adverse effects at high deposition rates, e.g. during snowmelt. In addition to the wet deposition of nitrate and ammonium, there is input to land and lake ecosystems through dry deposition of gaseous and particulate nitrogen species. The amount of this input is not well known, but there are data indicating that it can be substantially higher than wet deposition.

195. Acid deposition from the air into aquatic systems occurs today primarily through emissions of SO_2. Nitrogen and sulphur compounds lead to a rapid increase in acid concentration, above all during periods of snowmelt. Although broad, long-term reductions in SO_2 emissions are expected and have been recorded in some countries, the opposite trend is observed for NO_x emissions. It is to be expected that increases in water acidification owing to nitrogen compounds will occur.

196. The International Co-operative Programme for assessment and monitoring of acidification of rivers and lakes is expected to provide more complete information on dose-response relationships under different geographical conditions. The Co-operative Programme should also make it possible to identify long-term trends and variations in the chemistry and biota of aquatic ecosystems attributable to the impact of atmospheric pollution and subsequent acid deposition. The extent of present acidification of surface waters on a geographical scale for the different countries should be established, including numbers and sizes of lakes, number and lengths of streams. Suitable biological indicators should be presented for the biological monitoring.

197. Specific attention should be paid to the acidification of lakes and streams in nature reserves and national parks.

198. Water-quality models and dose-response functions for biota should be developed in such a way that they assist in formulating critical loads and, hence, assist in determining air pollution control policy. There is a need for further development of predictive models of surface water and groundwater acidification. In order to be able to predict the environmental consequences of acid deposition more accurately, thorough understanding is needed of geochemical processes and soil weathering under acidifying conditions. There is also a need to evaluate existing long-term reliable water chemistry records for sensitive waters in acidified areas.

199. Further studies should be made on short-term chemical fluctuations in running surface waters to understand the mechanisms and processes involved.

200. Further implications of acidification of surface waters should be studied, taking into account possible consequences for water management, including regeneration of acidified aquatic ecosystems as well as reversibility.

REFERENCES

1. E.C. Krug and others, "Appraisal of some current hypotheses describing acidification of watersheds", Journal of the Air Pollution Control Association, 35; (pp. 109-114, 1985).

2. H.M. Seip and S. Rustad, "Variations in surface water pH with changes in sulphur deposition", Water Air Soil Pollution, 21; (pp. 217-223, 1984).

3. A. Henriksen, "Changes in base cation concentrations due to freshwater acidification", Verhandlungen der Internationalen Vereinigung für Theoretische und Angewandte Limnologie, 22; (pp 692-698, 1984).

4. B.D. Lazerte, and P.J. Dillon, "Relative importance of anthropogenic versus natural sources of acidity in lakes and streams of central Ontario" (Canada), Canadian Journal of Fisheries and Aquatic Sciences, 41; (pp. 1664-1667, 1984).

5. L.M. Overrein, H.M. Seip, and A. Tollan, Acid precipitation - effects on forest and fish. (Final report of the SNSF project 1972-1980, 1980).

6. D. Fowler, "Transfer to terrestrial surface", Philosophical Transactions of the Royal Society, (London, 305 B; pp. 281-297, 1984).

7. M. Hauhs, "A model of ion transport through a forested catchment at Lange Bramke", Proceedings of workshop on mechanisms of ion transport in soils, (Zurich, Switzerland, 1985).

8. Research activity catalogue 1984/85 on air pollution and acidification in Sweden, (SNV PM 1917).

9. A.S. Lefohn, and G.D. Klock, "The possible importance of forest soil processes in defining surface water pH depressions", Journal of the Air Pollution Control Association, 35; (pp. 632-637, 1985).

10. I.M. Nazarov, and A.G. Rjabotchapko, Acid-forming substances in the atmosphere and their impact on vegetation ecosystem productivity, (1982).

11. J.O. Reuss, and D.W. Johnson, Effect of soil processes on the acidification of water by acid deposition, Journal of Environmental Quality, 14; (pp. 26-31, 1985).

12. B.J. Cosby and others, "Freshwater acidification from atmosphere deposition of sulfuric acid: A quantitative model. A method for estimating the time scales of catchment acidification", Environmental Science and Technology, (1985).

13. G. Wiklander, "Loss of nitrogen by leaching from fertile forest soils in southern Sweden", Journal of the Royal Swedish Academy of Agriculture and Forest Science, (122/5; pp. 311-317, 1983) (in Swedish, English summary).

14. E. Matzner and others, "Elementflüsse in Waldökosystemen im Solling - Daten-dokumentation", Göttinger Bodenkundl. Berichte 71, (pp. 1-267, 1982) (in German).

15. J.W.M. Rudd, and C.A. Kelly, "Microbial consumption of nitric and sulphuric acids in sediments of acidified lakes in four regions of the world", in: Abstracts, International symposium on acidic precipitation, (Muskoka (Canada), Abs. 221-222, 1985).

16. N. Christophersen, S. Rustad, and H.M. Seip, "Modelling streamwater chemistry with snowmelt", Philosophical Transactions of the Royal Society, London, 305 B, (pp. 427-439, 1984).

17. R.F. Wright, "Input-output budgets at Laugtfern, a small acidified lake in southern Norway", Hydrobiology, 101: (pp. 1-12, 1983).

18. R.F. Wright, "Acidification of freshwaters in Europe", Water Quality Bulletin, Vol.8, No.3, (pp. 137-142, 168 (1983)).

19. P. Grennfelt, and H. Hultberg, "Effects of nitrogen deposition on the acidification of terrestrial and aquatic ecosystems", Acidic precipitation, Proceedings of the International Symposium on Acidic Precipitation, Muskoka (Canada) 1985, (Part 1, pp. 845-964, Dordrecht 1986).

20. Environmental Protection Agency, "The Acidic Deposition Phenomenon and Its Effects", Critical Assessment Document" (EPA/600/8 - 85/001, 1985).

21. S.I. Nilsson, "The acidification sensitivity of Swedish forest soils" (SNV PM 1979, National Swedish Environment Protection Board, 1985).

22. Driscoll, "Aluminium speciation and equilibria in dilute acidic surface waters of the Adirondack region of New York State", in O.M. Bricker, ed. Acid precipitation, geological aspects, (Ann Arbor Science, Michigan, 55-75, 1983).

23. S.A. Meger, "Polluted precipitation and the geochronology of mercury deposition in lake sediments of northern Minnesota", Acidic Precipitation, Proceedings of the International Symposium on Acidic Precipitation (Muskoka (Canada 1985), Part 1, pp. 411-420, Dordrecht 1986).

24. A. Henriksen, O.K. Skogheim, and B.O. Rosseland, "Episodic changes in pH and aluminium-speciation kill fish in a Norwegian salmon river", Vatten 40: (pp. 255-260, 1984).

25. United Kingdom Acid Water Review Group, Interim Report, (1986).

26. J.H.D. Vangenechten, "Acidification in West-European lakes and physiological adaptation to acid stress in natural inhabitants of acid lakes", Water Quality Bulletin (8, pp. 150, 1983).

27. J. Fott, and others, Report on the present state of acidification of Black Lake, Dept. Hydrobiology, (Ch. Univ., Prague, Czechoslovakia, 1980).

28. R. Mosello and G. Tartari, Effects of acid precipitation on subalpine and alpine lakes, Water Quality Bulletin, (8, pp. 16, 1983).

29. C. Steinberg and others, Chemical sediment stratigraphy of four alpine lakes in Austria. Proc. Int. Symp. on Paleolimnology, (Carinthia, 1985).

30. H.W. Zöttl and others, "Chemismus von Schwarzwaldgewässern während der Schneeschmelze", Naturwissenschaften, 71 (2) (pp. 268-270, 1985) (in German).

31. Schoen, Wright and Krieter, "Gewässerversauerung in der Bundesrepublik Deutschland, Naturwissenschaften, 71 (2), (pp. 95-97, 1984) (in German).

32. G. Babiakova, and D. Bodis, "Snow cover as a source causing pollution of streams", Vod. Cas., 33, (pp. 468-485, 1985) (in Slovak, English summary).

33. D. Tuthar, "Current chemical and biological status of Yugoslav water-courses and lakes with respect to water acidification", Water Quality Bulletin, 3, (pp. 131).

34. I. Licsko, and K. Zotter, "Modification of the treatment technology of waters affected by acid rain", Water Quality Bulletin, 4, (pp. 190-195, 1985).

35. E. Plaza, "Chemical composition of precipitation in South Poland", Water Quality Bulletin, 4 (pp. 190-195, 1985).

36. R.W. Battarbee, "Diatom analysis and the acidification of lakes", Philosophical Transactions of the Royal Society, 305 B, (pp. 451-477, 1984).

37. R.J. Flower, and R.W. Battarbee, "Diatom evidence for recent acidification of two Scottish lochs", Nature, 305, (pp. 130-133, 1983).

38. J.G.M. Roelofs, "Impact of acidification and eutrophication on macrophyte communities in soft waters in the Netherlands", Aquatic Botany, 17, (pp. 139-155, 1983).

39. J.A.A.R. Schuurkes, "Atmospheric ammonia deposition and its role in the acidification and nitrogen enrichment of poorly buffered aquatic systems", (manuscript), (1986).

40. J. Kamari, "A quantitative assessment of lake acidification in Finland", Aqua Fennica, 15.1, (pp. 11-20, 1985).

41. R.F. Wright, "Rain project", Annual report for 1984 (Rep. 7/1985, NIVA, 1985).

42. SNV: Action plan against air pollution and acidification, (SNV PM 1862, Statens Naturvardsverk, 1984).

43. P.J. Dillon, "Chemical alterations of surface waters by acidic deposition in Canada", Water Quality Bulletin, 8: (pp. 127 ff, 1983).

44. J.M. Omerink and C.F. Powers, "Map supplement. Total alkalinity of surface waters: A national map", Annals of the association of American Geographers, 73, (pp. 133-136, 1983).

45. C.T. Driscoll, and R.M. Newton, "Chemical characteristics of Adirondack lakes", Environmental Science and Technology (Vol. 19 11, pp. 1018-1024, 1985).

46. D.E. Canfield, "Sensitivity of Florida Lakes to acidic precipitation", Water Resources Research, 19, (pp. 833-839, 1983).

47. R.B. Davis and others, Paleolimnological reconstruction of the effects of atmospheric deposition of acid and heavy metals on the chemistry and biology of lakes in New England and Norway, Hydrobiology, 103, (pp. 113-123, 1983).

48. K. Tolonen, and T. Jaakkola, "History of lake acidification and air pollution studies on sediments in south Finland", Annales Botanicae Fennici, 20, (pp. 57-78, 1983).

49. C. Steinberg, K. Arzet, and D. Krause-Dellin, Gewässerversauerung in der Bundesrepublik Deutschland im Lichte paläolimnologischer Studien. Naturwissenschaften 71, (pp. 631-633, 1984) (in German).

50. K. Arzet, D. Krause-Dellin, and C. Steinberg, "Acidification of selected lakes in the Federal Republic of Germany as reflected by subfossil diatoms, cladoceran remains and sediment chemistry", In: Smol, Battarbee, Davis, and Meriläinen, (eds.) Diatoms and Lake Acidity, (W. Junk, The Hague, 1986).

51. D. Krause-Dellin, and C. Steinberg, Evidence of lake acidification by a novel biological pH-meter. Environmental Technology Letters, 5, (pp. 403-406, 1984).

52. D. Charles, "Recent pH history of Big Moose Lake (Adirondack mountains) inferred from sediment diatom assemblages", International Association of Applied Limnology, (Lyon, France, August 1983).

53. R.A. Smith, and R.B. Alexander, "Evidence for acid precipitation induced trends in stream chemistry at hydrological bench-mark stations", U.S. Geological Survey Circular, (910, 1983).

54. E. Alasaarela, and P. Heinonen, "Alkalinity and chemical oxygen demand in some Finnish rivers during the periods 1911-1931 and 1962-1979", Publication of the Water Research Institute, (Finland, No. 57, 1985).

55. W.D. Watt, C.D. Scott, and W.S. Whilte, "Evidence of acidification of some Nova Scotia rivers and its impact on Atlantic salmon", Canadian Journal of Fisheries and Aquatic Science, 40, (pp. 462-473, 1983).

56. T.A. Clair, and P.H. Witfield, "Trends in pH, calcium, and sulphate of rivers in Atlantic Canada", Limnol. Oceanogr. 28, (pp. 160-165, 1983).

57. H.M. Seip, "Acid precipitation - literature review" (SI-810701-3, 1984).

58. H.M. Seip, "Acid precipitation - literature review" (SI-810701-4, 1985).

59. H. Hultberg, and P. Grennfelt, "Gardsjön Project: Lake acidification chemistry in catchment runoff, lake liming and microcatchment manipulations". In: Water, Air and Soil Pollution, 3011-2, (pp. 31-46, 1985).

60. M.D. Morgan, "Acidification of headwater streams in New Jersey Pinelands: A re-evaluation", Limnol. Oceanogr. 29, (pp. 1259-1266, 1984).

61. G. Andersson, and G. Gahnström, "Effects of pH on release and sorption of dissolved substances in sediment-water microcosms", Ecological Bulletin, 36, (1984).

62. B.I. Andersson, "Properties and chemical composition of surficial sediments in the acidified Lake Gardsjön", Ecological Bulletin, 37, (1984).

63. J.F.M. Geelen, and R.S.E.W. Leuven, Impact of acidification on phytoplankton and zooplankton communities, Experientia (in press, 1986).

64. O. Broberg, and G. Persson, External budgets for phosphorus, nitrogen and dissolved organic carbon for the acidified Lake Gardsjön, Arch. Hydrobiol. 99, (pp. 160-173, 1984).

65. P.A.E. Turner, and P.M. Stokes, The growth of filamentous algae isolated from acidic lakes grown over pH gradient 4.5 to 6.9, International Symposium on Acidic Precipitation, (Canada, Abs. 120, 1985) op. cit.

66. A. Melzer, and Rothmeyer, E. "Die Auswirkung der Versauerung beider Arberseen im Bayerischen Wald auf die Makrophytenvegetation", Bericht der Bayerischen Botanischen Gesellschaft, 54, (9-18), (in German).

67. A. Melzer, K. Held, and R. Harlacher, Die Makrophytenvegetation des Grossen Arbersees - neueste Ergebnisse. Bericht der Bayerischer Botanischen Gesellschaft, 56, (pp. 171-22, 1985) (in German).

68. A. Melzer, K. Held, and R. Harlacher, Die Makrophytenvegetation des Rachelsees im Bayerischen Wald. Bericht der Bayerischer Botanischen Gesellschaft, 56, (pp. 223-226, 1985) (in German).

69. J.G.M. Roelofs, "Effects of atmospheric sulphur and nitrogen, deposition on aquatic and terrestrial heathland vegetation", Experientia (in press, 1986).

70. Matthias: Der Einfluss der Wasserstoffionenkonzentrationen auf die Zusammensetzung von Bergbachbiozönosen, dargestellt an einigen Mittelgebirgsbächen des Kaufunger Waldes (Nordhessen/Südniedersachsen). Diss. Universität Kassel, 1982 (in German).

71. Marthaler, Limnologische Untersuchungen zum Einfluss von Düngemassnahmen auf die Fauna von Fliessgewässern. Dipl.-Arbeit, Universität Heidelberg, 1985 (in German).

72. W. Meinel, and S. Kleiner, in Wieting and others, Zum Einfluss saurer Niederschläge auf die Zoozönosen zweier Mittelgebirgsbäche im Kaufunger Wald. (Materialien 1/84, pp. 413-426, 1984) (in German).

73. W. Meinel, U. Matthias, and S. Zimmermann, Okophysiologische Untersuchungen zur Säuretoleranz von Gammarus fossarum (Koch). Arch. Hydrobiol. 104, (pp. 287-302, 1985, (in German).

74. M.P.D. Meijering, in Wieting and others, Die Verbreitung von Indikatorarten der Gattung Gammarus im Schlitzerland (Osthessen) in 1968 and 1982. (Materialien 1/84, pp. 96-105, 1984) (in German).

75. K. Clark, and R. Hall, "Responses of amphibian eggs and larvae to increased acidity and elevated alumnium levels in ponds in central Ontario", Canada. International symposium on acidic precipitation (Muskoka, Canada, Abs. 140-141, 1985) op. cit.

76. H.-J. Clausnitzer, "Gefährdung des Moorfrosches durch Versauerung der Laichgewässer", Naturschutz aktuell, No.2 (Kilda Verlag Greven, 1984) (in German).

77. R.S.E.W. Leuven and others, Effects of water acidification on the distribution pattern and the reproductive success of amphibians, Experientia (in press, 1986).

78. I.H. Sevaldrud, and I.P. Muniz, "Acid lakes and freshwater fishery in Norway". (SNSF Project IR77/80, 1980).

79. K. Johansson and P. Nyberg, Acidification of surface waters in Sweden. Effects and extracts. (SUNG 1981).

80. J.R.M. Kelso and others, in Proceedings, Effects of acid precipitation on ecological systems: Great Lakes region. (Michigan State University, 1981).

81. Henriksen and others, Rapport No. 97183, (NIVA, Oslo, 1983).

82. A. Jernelöv, "Effects of acid rain on mercury levels in fish", Water Quality Bulletin, 4, (pp. 204-205, 1985).

83. F. Eriksson and others, "Ecological effects of lime treatment of acidified lakes and rivers in Sweden", Hydrobiology, 101, (pp. 145-164, 1983).

84. M. Eriksson, "Acidification of lakes: Effects on water-birds in Sweden", Ambio 13, (pp. 260-262, 1984).

85. S.J. Ormerod, and R.W. Edwards, "Stream acidity in some areas of Wales in relation to historical trends in afforestation and the usage of agricultural limestone", Journal of Environmental Management 20, (pp. 189-197, 1985).

86. D.R. McNicol and others, Waterfowl and aquatic ecosystem acidification in northern Ontario, International symposium on acidic precipitation, (Muskoka, Canada, Abs. pp 144-145, 1985).

87. A. Henriksen, "Acidification of freshwaters - a large-scale titration", in: Ecological Impact of Acid Precipitation, D. Drablos, and A. Tollan, eds., (SNSF, pp. 68-74, 11180).

88. R.F. Wright, and A. Henriksen, "Restoration of Norwegian lakes by reduction in sulphur deposition", Nature, 305, (pp. 422-424, 1983).

89. S. Rustad and others, "Model for stream water chemistry of a tributary to Harp Lake, Ontario," Canadian Journal of Fisheries and Aquatic Science 43 (3) (pp. 625-633, 1986).

90. J. Kramer, and A. Tessier, "Acidification of aquatic systems - a critique of chemical approaches", Environmental Science and Technology, 16, (pp. 4606-4615, 1984).

91. W. Dickson, "Survey of acidification of Scandinavian freshwater systems, Water chemistry of (airborne) NO_x" in: T. Schneider, and L. Grant, eds. Air Pollution by Nitrogen Oxides, Elsevier, (pp. 555-565, 1982).

92. R.A. Goldstein and others, "Intergrated acidification study (ILWAS): a mechanistic ecosystem analysis", Philosophical Transactions of the Royal Society, (London, 305 B, pp. 409-425, 1984).

93. B.J. Cosby and others, "Modelling the effects of acid deposition: Assessment of a lumped parameter model of soil water and streamwater acidity", Water Resources Research, Vol. 21, (pp. 51-63, 1985).

94. R.F. Wright and others, "Interpretation of paleolimnological reconstructions using the MAGIC model of soil and water acidification", In: Water, Air and Soil Pollution, 301-2 (pp. 367-380, 1985).

95. R.F. Wright and others, Project RAIN: Changing acid deposition to whole catchments. Acidic Precipitation, Proceedings of the International Symposium on Acidic Precipitation, (Muskoka, Canada) Part 1, pp. 47-63, Dordrecht 1986).

96. H.M. Seip and others, "Model of sulphate concentration in a small stream in Harp Lake catchment, Ontario, Canada". Canadian Journal of Fisheries and Aquatic Science, 42, (pp. 927-937, 1985).

97. D.C.L. Lam and A.G. Bobba, "Modelling watershed runoff and basin acidification", in I. Johannson, ed. (pp. 205-215, 1985).

98. P.G. Whitehead and N. Christophersen, eds. SWAP, Modelling workshop report (Institute of Hydrology, United Kingdom, March 1985).

99. P.G. Whitehead, R. Neale and C. Neal, Predicting the effects of acid deposition on water quality. Second report on Contr (ENU-829-UK (H), IH, United Kingdom, 1985).

100. J.L. Schnoor and W. Stumm, "Acidification of aquatic and terrestrial systems", in: W. Stumm, ed. Chemical processes in lakes, (John Wiley, pp. 311-338, 1985).

101. J.C. Lin and J.L. Schnoor, An acid precipitation model for seepage lakes (1985).

102. J. Alcamo and others, "Integrated analysis of acidification in Europe", Journal of Environmental Management, 21, (pp. 47-61).

103. P. Kauppi and others, "Acidification of forest soils: A model for analyzing impacts of acidic deposition in Europe" (Version II, CP-85-27, IIASA, A-2361 Laxenburg, Austria, p. 28, 1985).

104. M. Posch and others, "Sensitivity analysis of a regional scale soil acidification model", Collaborative Paper, (CP-85-45, IIASA, 1985).

105. J. Kämäri, M. Posch, and L. Kauppi, "A model for analyzing lake water acidification on a large regional scale", Part 1: Model structure. (CP-85-48, IIASA, 1985).

106. P.R. Bloom, and D.F. Grigal, "Modelling soil response to acidic deposition in nonsulphate absorbing soils", Journal of Environmental Quality, 14, (pp. 489-495, 1985).

107. P. Kauppi and others, "Acidification of forest soils: model development and application for analyzing impacts of acidic deposition in Europe". (IIASA, CP-84-16, 41p, 1984).

108. J. Kämäri, "Methods for predicting freshwater acidification". (IIASA Working paper WP-84-27, 1985).

109. J. Kämäri, M. Posch, and L. Kauppi, "Development of a model analyzing surface water on a regional scale: Application to individual basins in southern Finland", in I. Johansson, ed. Hydrological and hydrochemical mechanisms and model approaches to acidification of ecosystems. (NHP rep. No.10, pp. 151-170, 1985).

110. R.W. Battarbee and others, "Lake acidification in Galloway: a paleoecological test of competing hypotheses", Nature, 314, (pp. 350-352, 1985).

111. H. Simola, K. Kenttämies, and O. Sandman, "Recent pH - history of some Finnish headwater and seepage lakes", Aqua Fennica, (15.2, 1985).

112. Characteristics of lakes in the Eastern United States, Vol. I-III, (EPA/600/4-86/007a, 1986).

Chapter 3

MECHANISMS AND EFFECTS OF WATER AND SOIL ACIDIFICATION ON STRUCTURES

201. Corrosion of materials in contact with soil and water is a complex problem. The corrosion rate and type of corrosion are determined by many factors. The effects of acidification including corrosion of materials have been known for a long time. So far, only the corrosion of structures above ground, caused mainly by sulphur pollutants, has been taken into account. This is hardly surprising, particularly as the influence of SO_2 on various materials has been the subject of many laboratory and field investigations. For historic and cultural monuments made of sandstone or limestone, the corrosive effect of a polluted atmosphere is obvious. The present state of knowledge of the effects of sulphur compounds in atmospheric corrosion was summarized in a report prepared within the framework of the Convention on Long-range Transboundary Air Pollution (1).

202. In view of the ongoing acidification of surface waters and soil and the risk of acidification of groundwater, the influence of corrosion on buried installations and installations in water also requires attention.

203. The corrosion effects of acid deposition may be subdivided as shown in the following diagram (2):

```
┌─────────────────────────────────────────┐
│     Corrosion owing to acidification      │
└─────────────────────────────────────────┘
```

```
┌──────────────────────────┐      ┌──────────────────────────┐
│ Atmospheric corrosion:    │      │ Soil and Water Corrosion: │
│ direct effects            │      │ indirect effects          │
│ (usually a local problem) │      │ (primarily a regional problem) │
└──────────────────────────┘      └──────────────────────────┘
```

204. Atmospheric corrosion can occur as a result of deposition of pollutants on the surface of materials. In most parts of the world this direct effect is usually restricted to areas near an emission source. Corrosion in soil and water is, on the other hand, an indirect effect of acidification; it may exert its influence on a regional scale. Long-distance transport of air pollutants may play an important role in this regard.

205. The purpose of the present report is to review briefly the data in the literature dealing with the effects of acid deposition on corrosion occurring in water and soil.

I. CORROSION IN SOIL

A. FACTORS AFFECTING CORROSION

206. Corrosion of metals in soil is an electrochemical process in which the dissolution of metal represents the anodic reaction. In most cases the cathodic reaction consists of reduction of oxygen. Reduction of sulphates to sulphide owing to bacterial activity may also be considered as a possible cathodic reaction under anaerobic conditions. With respect to corrosion, soil is a very complicated system. The corrosion rate and the type of corrosion are determined by numerous factors. These include mineral composition and grain size distribution of soil, moisture content, redox potential, electrical conductivity, pH, total acidity, chloride concentration, presence of sulphur and nitrogen compounds and microbiological activity.

Obviously, it is difficult to make a clear assessment of how acidification of soil affects the corrosion rate in so complicated a system.

B. CHANGES IN SOIL CHEMISTRY FROM ACIDIFICATION

207. With acid deposition, neutralization occurs through dissolution of carbonates and silicate minerals. In this process hydrogen ions are consumed and calcium, magnesium, potassium and sodium ions are released. The dissolution of basic minerals buffers soil against pH changes. If acidification continues, there will be a successive pH decline in the soil and soil water. Different areas are differently affected, depending both on the geological situation and on the deposition load of acidifying pollutants. In the absence of excess acid deposition, the pH of a soil will depend on its mineral content (buffering capacity) and microbial activity. Changes in pH as a result of acid deposition are a function of the same factors. In areas rich in limestone, for example, acid deposition will not have a major effect on soil pH. In other areas of Europe, particularly Scandinavia, the buffering capacity is much less (3). In a few studies, the pH of soil has been remeasured at the same point several decades later. These studies have revealed a substantial pH decline. For example, a recent Swedish investigation (4) found that acidification of forest soils in southern Sweden had led to a mean pH decrease of about 0.7 units over the past 35 years (Fig. 1). It should be stressed that pH decreases not only in the topsoil but through the whole soil profile investigated, which may have significance for the corrosion of buried structures.

C. EFFECT OF SOIL ACIDIFICATION ON CORROSION

208. A survey of the published literature shows that there are no systematic studies of the influence of soil pH on the corrosion rate of different metals. Nevertheless there have been investigations where soil acidification parameters were included among other variables. Some of these can be used for a preliminary evaluation.

Carbon steel, cast iron

209. Opinions are divided as to how far lowering of the soil pH will affect the corrosion rate of steel and cast iron. There is, however, general agreement that the corrosion rate increases when the pH falls below about 5. Some observers, taking the results of analyses of several thousand road culverts in California, found that lowering the pH value of soil increased the corrosion already at pH 6 (Fig. 2) or even at pH 8.5 (5-7). Others, however, considered that the corrosion rate of steel and iron did not increase until the pH fell below about 5 (8, 9). There seems to be general agreement that an increase of total acidity of soil increases the corrosion rates (Fig. 3) (9, 42). **Zinc, galvanized steel**

210. The extent of corrosion varies within wide limits; it may be very high in both acid and very alkaline soils (Fig. 4) (10). As the experimental evidence is limited, most authors rely on results of a major study from the United States of America (11). A Swedish study has also shown that at pH below 4 the corrosion rate of galvanized pole foundations was much faster than under less acidic conditions (12).

Copper

211. Copper and its alloys are corrosion resistant in most soils, as has been confirmed by practical experience with copper water pipes, for example. According to the aforementioned United States study, however, the corrosion rate of copper increased greatly if the pH fell below about 4 (13).

Lead

212. Lead has normally good corrosion resistance in soil owing to a protective coating of almost insoluble salts - principally lead carbonate and lead sulphate. Lead sulphate is stable even at very low pH values. Lead carbonate on the other hand, which is most often formed on lead surfaces in soil, dissolves on lowering of pH. Whether this entrains acceleration of corrosion may depend on the nature of the acids, but further work is required (c.f. para. 23, 67).

FIGURE 1

pH changes at different depth in forest soils of south Sweden between 1949 and 1984 (4)

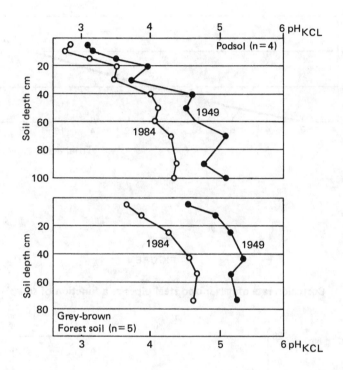

FIGURE 2

Time of perforation of a 2 mm galvanized steel culvert wall as a function of soil pH (7)

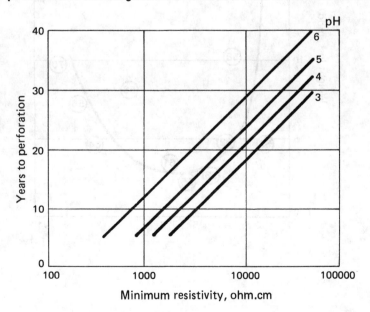

FIGURE 3

The corrosion rate of steel piping expressed by percentage of damage
to total length of pipe as a function of total soil acidity (9)

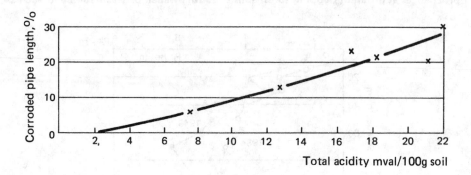

FIGURE 4

Corrosion rate of galvanized steel pipes as a function of soil pH (10)

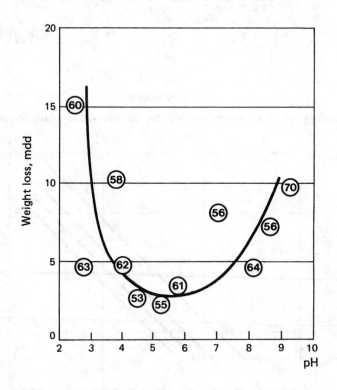

Concrete

213. Numerous factors such as water flow in the soil, soil chemistry and grade of concrete affect the corrosion rate of this material. An acidic sulphate solution may in principle attack concrete in two ways. Sulphate ions can penetrate the concrete and if present at relatively high concentration, react with lime and aluminates to form expansive reaction products which may crack the concrete. The risk of arriving at dangerous sulphate concentrations owing to acidification seems to be very low. Acidic water, however, weakens structures in the long run by leaching the cement phase out of concrete. The kind of acid seems also to be of importance; some authors believe sulphuric acid solutions do not cause corrosion until the pH falls to about 4 (14). On the other hand, the DIN Standard 4040/1968 of the Federal Republic of Germany contains the following classification of the effect of pH and sulphate concentration in groundwater on corrosivity of concrete:

	Degree of attack		
	Weak	Severe	Very severe
pH value	6.5 to 5.5	5.5 to 4.5	< 4.5
SO_4^{2-} mg/litre	200 to 600	600 to 3 000	> 3 000

214. An evaluation of the field performance of culverts in Ohio (United States) has shown that at pH values < 7.0 water pH has a significant detrimental effect on the durability of concrete pipe culverts (15). Dose/response functions expressing the rating for concrete have been discussed elsewhere. The variables with a significant effect in this respect are, in order of significance: water pH, incline, sediment depth and age. Cement mortar used as internal lining in iron and steel pipes has been proven to corrode much faster on lowering of the pH from 7 to 5 in a sulphuric acid solution (16).

Impregnated wood

215. Wood is often protected against decay by impregnation with an arsenic-based agent. The degree of fixation of the impregnant is reduced as pH drops (Fig. 5). Increased leaching into acid soil at pH < 5 may be expected to shorten the life of wooden materials, including telephone poles (17).

D. STRUCTURES WHICH MAY BE AFFECTED

216. Many essential structures are buried in soil. The following list, however incomplete, comprises a number of technically and economically important structures which may be potentially susceptible to soil acidification: water mains of steel and cast iron; conduits, connecting mains to houses, of galvanized steel, copper and lead; road culverts of galvanized steel and concrete; lead-sheathed telephone cables; oil and gasoline storage tanks; powerline tower foundations including stays; concrete sewers. A selection of these structures will be dealt with in greater detail to illustrate the extent and significance of the problem.

Powerline tower foundations and stays

217. Powerline pylons are usually of galvanized steel, sometimes set in direct contact with the soil, in other cases set in concrete. Steel or wooden poles are usually held by anchoring wires attached to galvanized steel stays, which are in turn fastened to underground anchors. The structures are built to high specifications and are generally considered to have a very long life. Owing to corrosion, however, the foundations and steel rods must be excavated and restored or replaced after only a few decades.

218. For example, some 40,000 towers with steel foundations have been installed in Sweden along with 1.5 million galvanized steel stays. The Swedish State Power Board (12) has examined around 50 tower foundations and found considerably higher than normal reductions of thickness of zinc at pH < 4. A shortened life owing to acidification implies considerable cost for repair or replacement. The extent of this problem varies from country to country and according to type of foundation.

Water mains

219. The total length of these pipelines is great. They are often made of cast iron and steel. The risk corrosion is highest on the outside where localized corrosion of the pitting type occurs. Their life varies greatly - sometimes 100-year old cast iron piplelines still function, elsewhere

FIGURE 5

Leaching of arsenic-based impregnant out of wood as a function of soil pH (17)

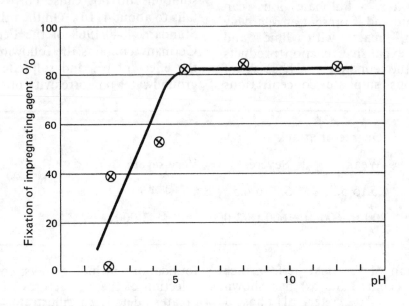

FIGURE 6

**Stream flow, pH and Ryznar Stability Index during an episode
of acid flow in a small water stream (21)**

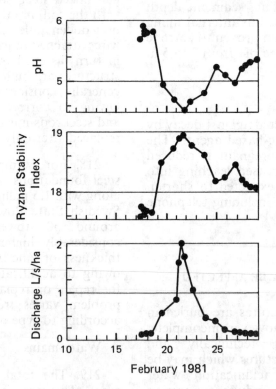

corrosion-damaged steel or cast iron pipes must be replaced after only 10 years. The very high total capital value represented by water mains indicates that a shortening of their useful life owing to corrosion caused by acidification may vastly increase the costs of damages.

Lead-sheathed cables

220. This type of telecommunication cable was extensively used during the 1950s and 1960s.

II. CORROSION IN WATER

A. FACTORS AFFECTING CORROSION

221. Corrosion of metals in water is mainly an electrochemical process in which the dissolution of metal represents the anodic reaction and reduction of oxygen dissolved in water forms the cathodic reaction. Only in extremely acidic waters with pH < 4 can evolution of hydrogen as a cathodic reaction begin to be of practical significance. The water composition is usually of decisive importance for the formation of protective layers on the metal surface. This in turn often affects the corrosion rate. Frequently the hydrogen carbonate ion (HCO_3^-) is the most important factor in the formation of protective layers. Water composition often determines whether certain types of local corrosion occur at all.

B. CHANGES OF WATER CHEMISTRY DURING ACIDIFICATION

222. On acidification of water, the HCO_3 concentration first falls, while sulphate (SO_4^{2-}) concentration and total hardness increase. As will be shown later, this development in itself can greatly increase the corrosivity of water towards certain metals. When the buffering capacity of the water has fallen to about 6 mg HCO_3^- per litre, the pH value becomes unstable and falls (2). This has been the case in surface waters in several locations both in southern Scandinavia as well as in the north-eastern United States of America and in Canada, where pH values between 4.5 and 5.5 occur.

223. For groundwater, reliable and representative time-series data are scarce for pH, sulphate, calcium, magnesium and hydrogen carbonate content. In south-eastern Sweden, in regions subject to acid deposition and where the

It is still not possible to predict the performance of these cables on the basis of corrosion data currently available. However, statistical data from the Swedish Telephone Company indicated that the frequency of corrosion damage to the cable network was highest in areas most sensitive to acidification of lakes and watercourses and where strong acidification of forest soil had also been observed (18).

bedrock consists of weather-resistant minerals, alkalinity decreases whereas the water's content of calcium, magnesium and sulphate increases in many wells in sand and gravel. In other areas, however, no direct relationship has been found between acid precipitation and decline in alkalinity (3).

C. EFFECTS OF WATER ACIDIFICATION ON CORROSION

224. The influence of water quality on corrosion of plumbing materials is a topic of great practical interest. The knowledge gained during many decades in connection with water treatment in waterworks may be used to evaluate the effect of water acidification on corrosion. The corrosion risk for the most important materials is discussed below.

Carbon steel, cast iron

225. The corrosivity of tapwater is usually considered to be related to the ability of the water to precipitate calcium carbonate which, together with rust, can form a tight and protective layer on steel surfaces. The ability of water to create protective layers is usually evaluated using so-called calcium carbonate saturation indices (19, 20). The most frequently used indices are the Langelier Index (L^s) and the Ryznar Index (R^s), which are defined as follows:

$$L^s = pH - pH^s$$

$$R^s = 2pH^s - pH$$

where pH is the actual pH value of the water and pH^s is the pH of the water at saturation with calcium carbonate. Waters with $L^s < 0$ resp $R^s > 7$ are considered as corrosive.

FIGURE 7

**Copper concentrations in tap water that
has been stagnant overnight in copper
pipes as a function of water pH (2)**

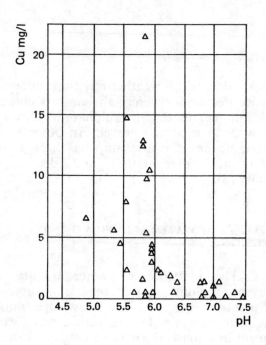

FIGURE 8

**The corrosion rate of zinc in 26 different tap
waters as a function of water pH (32)**

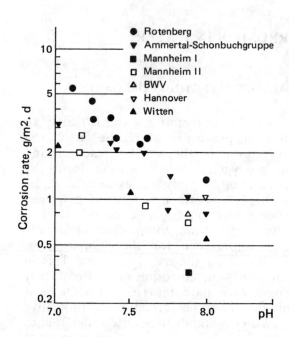

FIGURE 9

**Predicted metal loss for corrugated steel pipe
culverts as a function of water pH (49)**

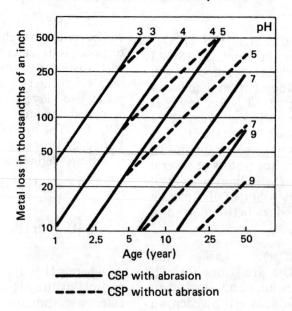

226. At water acidification the reading on L^s generally decreases whereas R^s increases and the waters become more aggressive. Quantitative relations between the saturation indices, water acidification and the corrosion rate of steel and zinc have not yet been subject to systematic investigations. However, recent research in two headwater streams which serve as the raw water supply for small water systems in Pennsylvania (United States of America) has shown that they were subject to adverse chemical changes during episodes of acid rain and snowmelt (21). The pH decreased as the stream flow increased, whereas corrosivity expressed as Ryzner Index (R^s) increased (see Fig. 6). In general it may be said that soft and acidic waters are corrosive. A soft water with low hydrogen carbonate concentration, however, gives no protective coating even if the pH value is raised.

Copper and copper materials

227. Copper is today in many countries a very important plumbing material. Acidification of water which results in lowering of the HCO_3^- content, an increase of the SO_4^{2-} content and a decline in pH can lead to different types of corrosion.

228. **Dissolution of copper (cuprosolvency)**, i.e. the concentration of dissolved copper salts in the water, increases with falling pH below a given critical level dependent on the temperature and composition of the water (22-24). A recent study in New Hampshire (United States of America) has shown that the value of the Langelier index (L^s) may also be used as a measure for the corrosivity of water towards copper (25). In time a protective layer of basic copper salts usually forms on pipes so that the normal copper concentration in tap water is around 0.1 to 0.8 mg/l. In areas with acidic water, considerably higher concentrations have been measured in the "first flush" of tap water that has been stagnant overnight (Fig. 7) (2). If the copper concentration in the water becomes too high, sanitary porcelain and textiles may be stained. Small amounts of copper - up to 1 or 2 mg per litre - are presumed to be not dangerous to health, at least not for healthy adults (26, 27). An effective countermeasure is to increase the pH value of the water to 8.0 or 8.5 (22).

229. **Pitting corrosion of type II**,[1] which occurs solely in hot-water installations, usually arises if the pH value is low (about 5 to 7), but not at pH < 7.4 (28). The phenomenon is usually encountered in water with relatively low concentration of hydrogen carbonate ions (< 100 mg HCO_3/l) and high concentration of sulphate ions. There is a great risk of pitting corrosion of type II if the HCO_3^-/SO_4^{2-} ratio is < 1. Acidification of the water thus increases the risk of this type of corrosion.

230. **Pitting corrosion of type III** is a type of corrosion which was first described in the Federal Republic of Germany (29). It was noticed in cold-water pipes at two places in Sweden in the 1970s and has recently been encountered at a third place. This type of corrosion was proven to be caused by the composition of the water. The damage arises in very soft and salt-deficient water; the pH of the treated water is high. All Swedish cases have occurred in acidified areas with extremely low concentrations of hydrogen carbonate in the raw water. In the first two affected localities, countermeasures have now been taken, consisting of raising the hydrogen carbonate concentration in the water. It should, if possible, be raised to around 100 mg HCO_3^-/l^{-1} (30).

231. **Erosion corrosion** occurs if the water flow in the pipe is highly pressurized so that protective layers are removed by the turbulence. Such attacks are not only more extensive but arise at lower flow rates if the pH of the water drops, for example, from 8 to 6.5 (31).

232. **Dezincification of brass** is a type of corrosion which may affect some grades of metal; it causes preferential dissolution of zinc from the material, leaving a porous residue of copper. The composition of water greatly affects its tendency to cause dezincification. Apart from oxygen, which must be present, there is a relation between the corrosivity of water and the hydrogen carbonate/chloride ratio. A low HCO_3^-/Cl ratio signifies that the water is corrosive from the dezincification aspect (32). A decreasing content of hydrogen carbonate owing to acidification may thus increase the risk of dezincification.

Galvanized steel

233. The corrosion resistance of galvanized steel water pipes depends on whether a protective layer of basic zinc carbonate (33-35) or a mixture of basic zinc carbonate with calcium carbonate (36) can be formed. A high hydrogen carbonate concentration in water contributes to the formation of such layers. In general, soft, acidic waters are corrosive, the pH of the water having a distinct effect on the protection offered by the layer (Fig. 8). At pH higher than 7.5, protective layers are usually formed and the corrosion rate

---

1 / Pitting corrosion of type I is not usually related to water quality.

of zinc is low. At pH below 7, on the other hand, the conditions for the formation of protective layers are seldom fulfilled.

Lead

234. Even if the use of lead for water pipes is decreasing, there are still areas in Europe as well as in Canada and the United States where lead is used in plumbing installations. Elevated lead concentrations have been found especially in tap water that has been stagnant overnight. The lower the pH and the concentration of salts, the more corrosive the water. The mean concentration in municipal water supplies in the United States of America is around 10 ug/litre. In some areas, e.g. in Boston (United States of America) lead concentrations higher than 50 ug/litre, have been noted, while elsewhere, where water is stored in lead-lined tanks, concentrations above 1 mg/l have occurred (37-41). Plumbosolvent water is sometimes considered to be related to a high humus concentration.

D. STRUCTURES WHICH MAY BE AFFECTED

235. Among the materials in contact with fresh water there are several which are of social significance. For most, present knowledge is not sufficient for an assessment of whether acidification has any practical relation to corrosion rates; when increased damage has been found, knowledge has usually not sufficed for quantification. Plumbing systems in buildings and road culverts will be used as an illustration of the problem.

Water pipes

236. The internal corrosion of water pipes caused by acid groundwater and surface waters is now evident. The corrosion of copper, galvanized steel, and lead pipes can be considerably increased by acidification of water supplies, with falling pH and hydrogen carbonate concentrations and rising sulphate concentrations. This is particularly manifest if the pH falls below about 7. As these conditions are well known, water quality requirements with respect to different materials have been set out in requirements and regulations.

237. The Swedish Building Standards may be mentioned among other examples (43). The municipal waterworks in Sweden treat water so that the pH is usually between 7 and 9. In areas with acidic water supplies, the hydrogen carbonate concentration as well as the pH should be raised by treatment. In houses receiving treated water from municipal waterworks it has probably not been of practical significance in most cases if the initial source of water has become more acidic, but it increases the costs of equipment and treatment chemicals in the waterworks.

238. It should, however, be pointed out that the corrosion risk in drinking water supplies is closely related to their management. Risks are generally greater the smaller the water supply. The greatest risk usually arises with small, privately owned supplies (44).

239. Among private water supplies in Sweden, serving some 1.1 million people, there may be a number in which acidification has increased the corrosivity of the water and caused costly corrosion damage. In such supplies, a possible counter-measure would be to raise the pH of the water, and preferably its hydrogen carbonate concentration as well, for eample, by installing an appropriate alkaline filter.

240. Also in the north-east of the United States of America and in Canada in areas with geological conditions similar to Scandinavia, the same problems are likely to be encountered. This may also apply to the numerous public water supplies which do not use corrosion control measures. In this connection the potential health hazard of corrosion products from the widespread use of lead piping should be mentioned (25). Shallow well water in the Adirondack Mountains (United States of America) was found to be particularly corrosive, suggesting the influence of acid precipitation (45).

241. In roof-catchment systems, which consist of a rain collector and a storage tank, corrosion problems arise in the plumbing installations in areas with acid precipitation. These systems are common in some regions in the eastern United States of America, where groundwater has been polluted by mining activity and public water supplies are not available. The bulk precipitation is today very corrosive as indicated by the Langelier Index (L'). The tap water samples from such systems plumbed with Pb-soldered copper pipes showed in 22 per cent of cases mean standing tap-water lead concentrations above 50 ug/litre and 28 per cent had Cu concentrations exceeding 1,000 ug/litre drinking water limit for copper (39).

242. In relation to water supplies, the corrosion of pump elements such as bowls and impellers of cast iron and bronze in groundwater should also be mentioned. The parameter of water pH was found to be the closest indicator of a material's lifetime (46).

243. With present knowledge it is not yet possible to quantify the effect of acid deposition on increased corrosion in plumbing systems. A study of the geographical distribution of corrosion damage to piping systems in buildings, and the costs, was made by the Swedish National Environmental Protection Board (47). It shows that the number of insurance claims for corrosion failures per 1,000 inhabitants annually was greatest, at pH < 4.15, in Krononberg County and that high figures were also usual in other areas likewise susceptible to acidification. The lowest figure, 0.84, was that for Uppsala County with calcareous soils.

Road culverts

244. Acidification of lakes and watercourses may also affect the corrosion rate of structures such as sluice gates, conduits in power stations or road culverts of galvanized steel and concrete. The documented evidence today is limited. An investigation of the corrosion of galvanized steel road culverts and of the pH of the ambient water and soil was made in Sweden at the end of the 1960s (48). It showed that the number of perforations and of severe corrosion attacks was considerably greater when the pH of the surface water was below 5.0 and 5.9, respectively, than when it was above 6.

245. Investigations in the United States of America have also revealed that corrosion of galvanized steel culverts in Ohio was considerably higher in waters with low pH (see Fig. 9) (7). The detrimental effect of water with pH lower than 7 on the durability of concrete pipe culverts has already been mentioned. Delamination of polymer coatings where the base culvert pipe is exposed is also accelerated by low pH water (49).

III. COST OF MATERIAL DAMAGE

246. No known systematic studies have been performed on the economic impact of increased corrosion rates owing to acidification of water and soil. Taking into account the very great quantities of the aforementioned structures and their high replacement cost, a reduction of their lifespan as a result of corrosion caused by acidification could give rise to substantial damage costs. Such damage cost assessments should be included in cost-benefit analysis for sulphur and nitrogen emission control. Assessments must identify the regions susceptible to acidification and also distinguish the component attributable to acid deposition, since other factors such as local agricultural practice may be significant.

IV. GAPS IN KNOWLEDGE

247. There are today major gaps in knowledge about the effect on structural materials owing to soil and water acidification. For corrosion in soil dose-response relations must be established to determine the acidification parameters in various types of soils and the corrosion rates for important construction materials. This may be achieved by both basic laboratory investigations and field studies as well as by systematic inspection of corrosion damage on existing structures. The influence of both sulphur and nitrogen pollutants should be taken into account in such investigations.

248. For corrosion in water, there is a need to quantify how acidified water - with changed pH, alkalinity and concentrations of sulphate, nitrate, heavy metals, organic substances, etc. - affects the corrosion rate of important engineering materials such as steel and cast iron tubes, galvanized steel tubes and concrete structures. Such studies should reveal dose-response relations for assessing the cost of corrosion damage. A deeper understanding of pitting corrosion of type III in copper tubes is also needed.

249. For both corrosion in soil and corrosion in water there is a need for input data for cost-benefit calculations, particularly the stock at risk, unsusceptible regions, and the effect of acid deposition on soil and water pH in these regions.

V. CONCLUSIONS AND RECOMMENDATIONS

250. Corrosion owing to acidification of water and soil is principally a regional problem and thus affected by long-range transport of air pollutants in susceptible areas.

251. Corrosion in soil is a complicated process and is determined by numerous factors. Dose-response relations between the corrosion rate and pH or acidity are still lacking. There are, however, indications that soil acidification may increase the risk of corrosion of important technical and construction materials.

252. The decrease of alkalinity and pH as well as the increase of sulphates owing to acidification increase the corrosivity of water to important engineering materials such as steel and cast iron, galvanized steel, copper, lead and concrete. For some of the related corrosion phenomena, criteria of occurrence are known. However, dose-response relations are still lacking.

253. Internal corrosion of water pipes of copper, lead, and galvanized steel has become a great problem today in areas sensitive to acidification, especially where there is no corrosion control of water supplies. Where lead pipes or lead-containing solder are used, elevated lead content in tap water also creates a potential health hazard. Increased corrosion of road culverts can also be expected in acidified areas.

254. Among buried structures identified as potentially susceptible to soil acidification figure distribution mains and house connections, certain types of foundations, the footings of powerline pylons and similar towers, including stays, lead-sheathed telephone cables and underground fuel tanks, etc.

255. There are major gaps of knowledge in this field. The need for research comprises, among other aspects, assessment of the effect of acid deposition on soil and water pH in different regions, studies of the mechanisms of corrosion from acidification by sulphur and nitrogen compounds, establishment of dose-response relations and cost estimates for damage from acidification.

REFERENCES

1. Airborne Sulphur Pollution: Effects and Control, Air Pollution Studies No.1 (ECE/EB.AIR/2, United Nations publication, New York, Sales No. E.84.II.E.8, 1984).

2. Acidification today and tomorrow (Ministry of Agriculture, Stockholm, 1982).

3. Effects of Sulphur Compounds and other Air Pollutants on Soil and Groundwater, L. Norberg, ed. Report 3002 (National Swedish Environmental Protection Board, Stockholm, 1985).

4. G. Tyler and others, Soil Acidification and Metal Solubility in Forests of south Sweden (NATO Advanced Research Institute Workshop, Toronto, Canada, May 1985).

5. R.F. Stratfull, Corrosion 17 (p. 493, 1961).

6. O. Arrhenius, "The Corrosion of Steel in Soil", Tekniska Skrifter No. 159 (Teknisk Tidskrifts Förlag, Stockholm, 1965).

7. Galvanized Steel in Soils, AHDGA Market News (Winter 1982).

8. E. Schaschl and G.S. Marsh, Materials Protection 2(p. 8, 1963).

9. H. Steinrath, Gas- und Wasserfach 106(p. 1361, 1965) (in German).

10. J. Marcovic and N. Plavsic, Werkstoffe und Korrosion 2 (p.85, 1961) (in German).

11. M. Romanoff, "Underground Corrosion", United States National Bureau of Standards Circular 579 (United States Government Printing Office, Washington, 1975).

12. V. Victor, Korrosion och Ytskydd 11(p. 40, 1970) (in Swedish).

13. J. Marcovic, M. Sevidic and L. Rubinic, Werkstoffe und Korrosion 2(p. 87, 1960) (in German).

14. R.B. Andersson, International Conference on Groundwater and Man (Sydney, Australia, 1983).

15. J.O. Hurd, Symposium on Durability of Culverts and Storm Drains, (Transportation Research Board, National Research Council, Washington DC, pp 40-44, 1984).

16. Heinrich and others, 3R International 17, Heft 7 July 1978.

17. W. Dickson, Swedish Environmental Protection Board (Stockholm, Sweden, personal communication).

18. B. Sandberg, Corrosion of underground structures due to acidification of the soil. KHM Teknisk Rapport 43 (Statens Vattenfallsverk, Vällingby, 1982) (in Swedish).

19. W. Langelier, Journal AWWA 28(p. 1500, 1936).

20. J.W. Ryznar, Journal AWWA 36(p. 472, 1944).

21. R.T. Liebfried, W.E. Sharp, D.R. De Walle, Journal AWWA 76(p. 50, 1984).

22. E. Mattsson, Proceedings of the Australasian Corrosion Association Silver Jubilee Conference (Adelaide, Australia, pp. C-7-1, 1980).

23. P.T. Gilbert, Proc. Soc. Water Treatment and Examination 15(p. 165, 1966).

24. S.A. Cox, B.I. Dillon, Proceedings of the Australasian Corrosion Association Silver Jubilee Conference (Adelaide, pp C-4-1, 1980).

25. F.B. Taylor, G.E. Symons, Journal AWWA 76(p. 35, 1984).

26. H.J.Holtmeier, M. Kuhn, I. Merz, Metall 32 (p. 1157, 1978) (in German).

27. Literature Review of the Effects of Copper Ingestion on Human Beings. (BNF Metals Technology Centre, Misc. Publ, No.591, 1978).

28. E. Mattsson and A-M Fredriksson, British Corrosion Journal 3(p. 246, 1968).

29. O. von Franqué, Werkstoffe und Korrosion 26 (pp. 255-258, 1975) (in German).

30. M. Linder, E-K. Lindman, Proc. 9th Scand. Corr. Congress. Korrosionscentralen, Copenhagen, 1983.

31. L. Knutsson, E. Mattsson, B.E. Ramberg, British Corrosion Journal 7 (p. 208, 1972).

32. M. Linder, "The influence of water composition on the corrosion of different materials in water pipes", KI Rapport 1984:1 (Swedish Corrosion Institute, Stockholm, 1984) (in Swedish).

33. DIN 50930, Teil 3, Korrosionverhalten von metallischen Werkstoffen gegenüber Wasser (Feuerverzinkte Eisenwerkstoffe, 1976) (in German).

34. J. Hissel, Corrosion of Galvanized Steel Related to Water Composition. Proc. of 12th Int. Galvanizing Conference (pp. 85, Paris, 1979).

35. A. Bächle et al, Werkstoffe und Korrosion 32(1981)435 (in German).

36. G. Butler and H.C.K. Ison, Corrosion and its Prevention in Waters (Leonard Hill, London, 1966).

37. The Health and Environmental Impact of Lead and an Assessment of Needs for Limitations (United States Environmental Protection Agency, Washington, DC, 1979).

38. A.D. Beattie and others, British Medical Journal 2(p. 490, 1972).

39. E.S. Young and W. Sharp, Journal of Environmental Quality 13(1984)38. 40. W.N. Richards and M.R. Moore, Journal AWWA 76(p. 60, 1984).

41. C.S. Wong and P. Berrang, Bulletin of Environmental Contamination & Toxicology 15(p. 530, 1976).

42. K.H. Logan, Underground Corrosion. United States National Bureau of Standards Circular 579 (United States Government Printing Office, Washington, 1975).

43. Svensk Byggnorm (Swedish Building Standard) SBN 1980 (Statens Planverks författningssamling 1980:1) (in Swedish).

44. T.W. Clarkson and others, The Acidic Deposition Phenomenon and its Effects. EPA-600/8-83-01 B (United States Environmental Protection Agency, Washington, DC).

45. J.S. Reed and J.C. Henningson, Journal AWWA 76(p. 60, 1984).

46. G.J. Kelly, Proc. of Int. Conference on Ground Water and Man (1983).

47. E. Levlin, Corrosion damage on water pipes in buildings and soil due to acidification. SNV PM 1978 (National Environmental Protection Board, Stockholm) (in Swedish).

48. O. Arrhenius, Corrosion of road culverts of corrugated galvanized steel plate. Bulletin No. 66 (Swedish Corrosion Institute, Stockholm, 1966) (in Swedish).

49. J.O. Hurd. Symposium on Durability of Culverts and Storm Drains. Transportation Research Board (National Research Council, Washington DC, pp. 35-40, 1984).

Chapter 4

DAMAGE TO MATERIALS BY AIR POLLUTION: METHODS FOR ASSESSING STOCK-AT-RISK

I. MATERIALS INVENTORY

256. The determination of the amount of exposed materials susceptible to acid deposition is known as the materials inventory. There are basically two approaches to preparing a materials inventory. One is a true inventory, or census, which enumerates each building/structure and provides an individual record describing its major features. In this approach, damage is calculated for each building and then added up to provide the total damage for all buildings in the inventory. The other approach begins by first totalling all the areas of exposed materials. The damage is then computed by using an average damage-function applied to this total. In this second approach, the identity of individual buildings is lost, and the materials inventory is characterized by a probability function for the exposed material per unit area of land. The second approach is sometimes called a materials distribution assessment to emphasize that it is not a true building-by-building inventory.

257. The probability distribution is developed statistically from data taken at a small number of field sites. The data is collected by actually going to a selected location, recording the materials found there on a number of sample buildings, and from this data calculating average values of materials per unit of land that can be used to project from the sampled area to the entire region of interest.

258. However, these average values are crude approximations, since it is unlikely that the materials distribution remains uniform and constant over any significant geographic area. Typically, as one goes from a central business district outward through suburban residential areas to rural countryside the relative percentage of materials varies. In the high-rise business districts materials such as granite, limestone and aluminium may appear often, but become infrequent in suburban areas where brick and painted wood may dominate. It is also evident that the absolute amount of material per unit of land decreases as one goes from high density urban areas to rural areas where buildings are widely separated.

259. Therefore, in order to improve the accuracy of the materials distribution estimates, variations in materials use could be predicted by characteristic variables such as population density or land-use classification. A materials distribution based on the use of predicator variables is known as a materials distribution model.

260. The choice between these two approaches depends upon the structure of the intended cost-benefit analysis and usually comes down to the dichotomy between cultural and ordinary architecture. In the cultural case, each building is essentially unique, even though the materials involved may be similar. For example, the Palais des Nations in Geneva uses limestone in a very different way from the Cathedral of St. Peter. Moreover, if the number of buildings of cultural interest is relatively limited, this should make a building-by-building enumeration feasible. Conversely, where the total number of ordinary buildings is on the order of tens of millions, many of them may be practically indistinquishable from one another so that a statistical distribution approach might be justified.

A. HISTORICAL BUILDINGS AND CULTURAL MONUMENTS

261. Historical buildings and cultural monuments are characterized by the desire to

71

maintain them in at least their present state almost indefinitely with a minimum of materials replacement or modification. Since their economic value cannot be derived in the same way as for ordinary modern buildings, they have to be treated separately in any economic assessment.

262. This strict definition is applicable only to a limited number of buildings and monuments of exceptional historical or cultural interest.

263. It is often assumed that a materials inventory for such buildings and monuments can be derived from existing data bases, such as registers of historic buildings. However, the amount of documentation for each building can vary enormously. For example, in France, a national computerized data base has been set up that contains an extensive description of each historic building including its materials and its present condition (Repellin, 1983). In other countries, the only information available for each building may be the location and date of construction.

264. At a workshop held in the United Kingdom in 1986, attention was drawn to the need for countries to provide certain data on their most important buildings and monuments in a common form. A main additional requirement in this context is an assessment of the visual appearance on an agreed scale of ranking since changes in visual appearance affect the aesthetic value of the buildings and monuments. Surface appearance is often a first indicator of degradation of structural integrity.

265. It was recognized that the categorization of classes of visual appearance will depend in part on the differing national practices for maintenance and repair of historic buildings and cultural monuments. In this connection, reference should be made to the standards developed by the International Institute for Conservation of Historic and Artistic Works. It was also recommended that, at the present time, other historical buildings and cultural monuments not covered by the above definition should be analysed using similar methodology as that used for modern buildings.

266. It is often assumed that a materials inventory of historical buildings and cultural monuments can be derived from existing data bases such as registers of historic buildings. In theory, this is possible, since registers exist for many places, but there are several reasons why this is usually not practical. One reason is that

the term cultural property is not restricted to buildings of architectural significance. Most registers include archaeological sites, which may be nothing more than traces of walls in the soil. Also, a building may be included on a register because of its historic associations rather than its aesthetic importance. For example, Ford's Theater in Washington, DC (United States of America) is a national landmark because it was where Abraham Lincoln was shot, rather than having any outstanding architectural merit.

267. Another problem with registers of cultural property is that it is sometimes difficult to decide how to count the structures that may be found at a particular site. For instance, Versailles is usually thought of as a single landmark, but it is actually a collection of buildings along with a large number of outdoor statues. Many cities contain old towns or historic districts that consist of groupings of buildings.

268. Finally, in addition, the registered buildings are usually only a subset of all the buildings that could be listed. This is owing in part to limited resources constraining the rate at which buildings can be documented. To complicate matters, the number of eligible structures is ever increasing, as more and more existing buildings reach the minimum age for listing, and as the definition of significant architecture has been broadened to include industrial architecture, vernacular architecture, etc. (Wills, 1983). Moreover, in many cases, listing on a register is voluntary, and the owners of private property may decline to nominate their buildings (Thurber, 1985).

269. In summary, although a materials inventory approach is indicated for cultural property, it is not clear that the necessary data can be obtained from existing registers. Moreover, it would be difficult to utilize these registers for comparative international studies of materials damage because of differences in definitions and in documentation. The European Economic Community is presently sponsoring a pilot study on the problem through a research grant to the United Kingdom and Federal Republic of Germany to make an inventory of several historic districts in those two countries. However, it may ultimately be necessary to resort to a statistical approach to the estimation of cultural property. Some research along these lines is being undertaken by the National Park Service in the United States (Sherwood, 1984).

B. MODERN BUILDINGS/STRUCTURES

270. For ordinary buildings/structures, only a materials distribution model is feasible. In this case it is necessary to specify the acceptable levels of uncertainty in the means and variances because these may well determine the minimum number of buildings that must be included in a ground survey. The ground survey work, which involves sending teams of researchers into the field to record the data, is the most expensive part of the process. It is possible to maximize the accuracy of the results for a given number of samples through advanced statistical sampling techniques. The most straightforward method of ground surveys, which involves a simple random selection of sample ground area units in a given geographic area, is usually not the most efficient way because some of the sample points may turn out to be vacant land (EPRI, 1983). It is now generally accepted that the sampling design should be based on sampling a specified number of buildings rather than simply units of land.

271. To minimize the amount of unnecessary survey effort, stratified sampling schemes have been developed. These take advantage of the fact that the building types in urban areas generally occur in distinct patterns, such as commercial districts, single-family residential neighbourhoods, industrial zones, etc.   An efficient sampling scheme would allocate a certain number of buildings to be recorded in each of these areas. Otherwise, all the data may come from only one type of neighbourhood, thereby skewing the materials distribution model if it is not representative.

272. Similar approaches have been developed to identify and characterize various areas. The Commission of the European Economic Community and the United Kingdom Department of the Environment have been supporting a programme of work on both historical and modern buildings (ECOTECH, 1985) which has, among other objectives, the development of improved stock-at-risk inventory methodologies and the development of models representing building damage mechanisms. The work on building damage mechanisms is concerned with deriving a baseline of current knowledge on the various causes of building damage and assessing the relative importance of the effects of air pollutants to this damage. Such data will be useful in interpreting the estimates of damage costs.

273.   With   regard   to   stock-at-risk methodologies, the work has concentrated on developing a methodology which allows an estimate to be obtained of the total quantity (surface area) of exposed materials in buildings and also re-defines stock-at-risk in terms of building components which may be of greater use in damage cost estimation. The methodology involves deriving so-called "identikits" for different types of buildings.   The stock of buildings is separated into 15 categories of buildings ranging from single-family units to industrial premises.   Each building is specified by the following: number of storeys, floor space index, basal surface index, number of buildings per square kilometre, and age of building.

274. It is possible to estimate the area of the external envelope (walls, windows, roof) using standard quantity surveyor techniques and field measurements for each building type.   Field surveys are then used to determine the typical materials used in the construction of the external envelope (e.g. galvanized steel, concrete, brick, cement roof tiles).   The grossing-up is based upon land-use maps which show the different building-type   patterns   to   determine   the proportion of each kilometre grid square occupied by a particular building type.   This methodology has been applied in four European cities : Birmingham and Lincoln in the United Kingdom, and Dortmund and Cologne in the Federal Republic of Germany.   The project has also re-defined the stock-at-risk to be a function of buildings and building components (e.g. gutterings, window units, window sills, etc.).   An inventory   of   building   components   using "identikits" has also been obtained for the above-mentioned cities.   Using published cost estimates for remedial works, it is possible to obtain a damage cost estimate which may be more representative of the costs of treatment since it takes into account the function of the components in a building.

275. In the Nordic countries, a joint programme on assessing reduced corrosion damage resulting from reduced SO_2 emissions is currently being carried out.   The Swedish and Norwegian projects deal primarily with the question of stock-at-risk inventory, while in Finland the work is concentrated on dose-response relationships.   Two separate inventories are being carried out using identical sampling methods and building survey protocols. The Swedish project covers the materials-at-risk in Stockholm, mostly affected by a large number of local sources.   The Norwegian survey concentrates on smaller industrial towns affected by a few, large industrial sources of SO_2. If the methodology proves successful, it will be applied also to other Nordic cities.

276. The sampling method and building survey protocols are based partly on earlier studies by the Swedish Institute for Building Research which examined the condition of the building stock in Sweden. The sampling procedure is based upon the classification of communities into four categories of annual mean winter SO_2 concentration. Buildings are grouped into nine categories: single-family houses up to 1920, 1921-1960, 1961-1984; apartments up to 1920, 1921-1960, 1961-1984; farmhouses; industry, and shops and offices. Data on real estate taxation is used to identify target sample buildings which are to be surveyed. A total of 400 buildings will be surveyed. The building survey includes obtaining information on the local environment, design data, materials inventory and a description of the condition of the building. The materials for inventory are subdivided into nine building components: foundation, walls, windows, doors, balconies, roof, gutters, ancillary buildings (e.g. garage); once the studies have been performed they will then form the basis of a model to be used in other Nordic cities.

277. One study also includes an assessment of the condition of exposed materials. Both the substrate material and the surface finish are classified into one of three classes: unaffected, slightly affected, and badly affected. In Canada, Leman et al. (1985) classified urban land into "urban textures" which were determined from aerial photographs.

278. As an alternative to manual methods, in the United States an automated procedure has been developed to classify urban areas into sampling frames using information in national computer data bases such as the Population Census (Rosenfield, 1984). Additional work in this field has been made for the United States Environmental Protection Agency. Besides the acid deposition programme, the sampling-frame approach has been applied to other problems such as predicting the number of buildings that may contain hazardous levels of asbestos (Strenio et al., 1984).

279. The amount of data to be recorded for each building also needs to be considered in light of conserving effort in the field and in computer time. The tendency is to collect as much detailed information as possible, but this must be weighed against the increased costs as well as the degradation of accuracy owing to sparse data in each category (Merry et al., 1985). For example, it is desirable to record each type of stone found in the survey: marble, limestone, granite, arkose sandstone, arenaceous sandstone, etc. However, if only one or two of each type is found in the

survey, the uncertainty in the materials distribution model will be high. Consequently, the categories are often collapsed into one general stone category, with better accuracy. On the other hand, the estimates of damage based on this aggregation become less meaningful because limestone has a very different damage function from granite. Inventories will also need to include data on more recent building materials. In this category are, for example, laminates and composites.

280. Once the data have been collected, the distribution model is developed through multiple correlations against predictor variables. The underlying concept is that there is some rational process governing the choice of materials for buildings. Factors that would presumably be involved include climate, structural strength, intended use, availability of materials, construction techniques and architectural style. It is evident that wooden framing cannot be used to build ten-storey buildings, so that this material would not be found in high-density urban areas. Building stone and wood are not very abundant in the Netherlands, but brick clay is, and consequently there is a high proportion of brick buildings in that country. Reinforced concrete construction is a recent innovation, and thus this material would not be found in a region of older buildings. There are characteristic materials associated with an Italia villa, a Bavarian farmhouse or a London townhouse, that reflect a combination of climate, available materials and local architectural tradition. All this implies that the distribution of materials is not simply a random process dependent upon the whims of individual builders.

281. The problem is to find quantitative measures and data that can predict these processes in a mathematical model. Population density and housing distribution are readily obtained from censuses. Land use data are now becoming available, and possible construction techniques can be inferred from building age. It is more difficult to quantify the distribution of an architectural style, although intuitively this would seem to be important. As noted by Matero and Teutonico (1982), the period from 1850 to 1890 in the architecture of New York City is known as the "brown decades" because brownstone was the principal building material. Bastian (1980) has reviewed the correlations between characteristic urban house types and such factors as building age, and social status and ethnic composition of neighbourhoods. Novak et al. (1984) have suggested that regional preferences can be inferred from current construction statistics.

282. The work of the International Organization for Standardization (ISO) through its working group ISO/156/WG.4 on classification of the corrosivity of the atmosphere will be beneficial for the studies on building-material assessment and their performance. The draft standard defines corrosivity categories using the most significant environmental factors (e.g. time of wetness, concentration of SO_2 and other air pollutants) by the observed corrosion losses for the four basic metals (steel, zinc, copper, aluminium) after the first year of exposure.

283. The guideline corrosion values will be useful for engineers in assessing the corrosion rates for different materials and for an optimal choice of materials and surface protection. The guideline values are based on the results of a large number of field exposure programmes in different climatic regions as well as practical knowledge gained by experts working in this field. The classification system is set out in four ISO Draft Proposals, namely ISO DP 9223, 9224, 9225 and 9226. It is envisaged that the classification will be extended to other materials as well. This methodology has been used as a basis for calculating the potential life of steel structures and has been applied to damage cost estimation for steel components in the city of

Prague (Czechoslovakia) and in industrial areas of Czechoslovakia.

284. Finally, there is the possibility of specialized inventories. The location of bridges and electrical transmission towers is a matter of record (Torpey and Lipfert, 1984). Building-by-building records are often available for central business districts (Lipfert and Torpey, 1984). In these situations, separate inventories may be more useful than trying to predict materials distribution from a general model.

285. Leman et al. (1985) have successfully characterized materials distribution in urban Toronto working mainly from information available in architects' offices. In many cases, the data recording was simplified because of the modular construction of most high rise buildings. Only in low-rise residential areas was it found necessary to conduct ground surveys. 286. Preliminary results from the United States programme indicate that in the north-east region, the area most at risk from acid precipitation, the predominant materials are painted wood and brick. The percentage of each shows some regional differences (Merry et al., 1985). An inventory of Los Angeles, on the west coast, found that the largest share of buildings was constructed of painted stucco (TRC, 1985), a material that has only minor use in the north-east.

II. SERVICE LIFE FACTORS

287. In order to put into perspective the physical damages computed from the deposition rates and the damage functions, it is necessary to convert them into service life changes. This is a way of providing a common yardstick for comparing damages to different materials, or to the same material used in different applications. For example, the loss of a given thickness of zinc coating from a chain link fence may have very different implications for the service life of that fence, compared to the consequences of the same amount of loss from a galvanized gutter. Similarly the loss of two or three millimetres of marble veneer on the side of a building may be imperceptible, while the loss of the same amount of material from the carved features of a statue may significantly impair its artistic value. This also provides a method of comparing damage among materials. While it would be difficult to relate the loss of so much zinc on a fence to the loss of paint from a wall, it is possible to compare the changes in service lives.

288. It must be emphasized that corrosion damage is not always the factor which determines service life. For example, repainting may be done for other commercial reasons. A building may be torn down when the cash flow it generates becomes less than the alternative of constructing a new building that will yield a greater return on the owner's investment. When corrosion is life-limiting, there are a number of ways in which the service life of a component can be determined (Taylor, 1968). For example, the loss of a cross-section of a bridge member through corrosion may reach a critical point where the member can no longer support a specified loading, resulting in catastrophic failure. A gutter may be corroded to a point where it begins to leak. However, it is not always possible to find such unambiguous criteria to mark the end of service life.

289. The concept of a critical damage level has been employed to define a level at which a material no longer performs its function (McCarthy et al., 1984). For example, Haynie

et al., (1976) assumed that for galvanized steel with a zinc coating of 25 microns, that failure occurs after the full 25 microns is removed. Thus, the critical damage factor in this case is a depth measurement.

290. In reality, however, the depth of a coating is not measured. Instead, the decision to replace materials is based on a visual observation of the amount of surface area that shows rusting or blistering. This can be as low as O.25 per cent of the area. (ASTM, 1983). Thus although the damage function estimates damage in one-dimensional terms (depth), the actual critical damage level is a two-dimensional problem (area). The appearance of individual blisters is a random process depending upon the location of random defects in the original coating. This becomes a problem in spatial statistics (Martin and McKnight, 1985a). The time until failure itself under these conditions is described by another type of probability distribution known as the Weibull distribution (Martin and McKnight, 1985b).

291. The implication is that the decision to replace a component is often taken before the time until failure predicted by a one-dimensional model. This emphasizes the need to develop damage functions for estimating damage in a two-dimensional sense, especially for surface coatings.

292. Manufacturers will sometimes specify recommended lives for certain components, charging a higher price for those designed to last longer, but there are few data available on actual performance relative to these specifications. Moreover, recommended lifetimes can differ for rural as opposed to industrial or urban locations, which suggests that some effects of pollution may already have been taken into account. Similarly, some professional societies may recommend criteria for determining the end of useful life. For example, the several indices of paint deterioration have been suggested using combinations of such measurements as erosion, chalking and rust spot areas. Finally, service lives may be specified arbitrarily in order to calculate depreciation. These economic service lives may bear little resemblance to actual physical service lives. Nevertheless, a compilation of these recommended service lives is a necessary first step to characterizing the situation.

293. Moreover, the actual service life may differ significantly from the recommended one for a number of reasons. The indifference of property owners or lack of funds for maintenance are two obvious ones. McCarthy et al. (1984) have introduced the concept of prevailing practice to describe the actual service lives as opposed to the nominal ones. However, to establish prevailing practice would require surveys of actual owner experiences or intentions. This has yet to be done.

294. One reason for distinguishing cultural properties from ordinary architecture is that the former are expected to have essentially infinite lifetimes. While this goal may be possible for certain objects kept in controlled museum environments, it may be unrealistic for materials used in architectural applications which are exposed to weathering processes. For example, much of New York City's nineteenth century architecture was built using brownstone of a particular type proven to be inherently prone to deterioration (Matero and Teutonico, 1982).

III. ECONOMIC VALUATION

295. The result of developing damage functions, service lives and materials inventories is an estimate of the physical damages caused by air pollution. However, the ultimate goal is to place an economic value on these damages in order to weigh them against the costs of control. It is customary to refer to the economic value of the reduction in damages due to a specified amount of air pollution control as the "benefit". In theory, if the benefits of a control option exceed its cost, then one would be justified in implementing that option. In practice, this type of explicit tradeoff is rarely achieved because of the problems in placing an economic value on the damages.

A. HISTORIC BUILDINGS AND CULTURAL MONUMENTS

296. The problem of estimating the benefits of reduced air pollution damage to ordinary buildings can at least be defined using conventional economic theory. The estimation of benefits involving cultural property becomes much more controversial. At one extreme, there are those who claim that such monuments are irreplacable and priceless. At the other extreme are those who maintain that the benefits are the same life-cycle costs as for ordinary buildings, with the possible addition of a premium paid for

special materials and workmanship (Barnes et al., 1983).

297. The problem can be redefined in terms of market versus non-market considerations. A market exists for ordinary building purchase, maintenance and repair, which provides some estimate of the prices the public is willing to pay. However, there is no similar market for historic or cultural buildings, many of which are either under public ownership or restrictive legislation. Moreover, there are parties interested in the protection of these monuments who do not enter into the direct transactions. Fund-raising activities on behalf of threatened monuments such as the Statue of Liberty or the City of Venice have shown that people are willing to pay to preserve these structures even when they may never be able to see them in person. This creates a value for such monuments beyond that measured by ordinary life-cycle costing. These extraordinary non-market values are given such names as options value, legacy value, etc., and much effort has gone into the development of surrogate market techniques to quantify them, both for buildings and for natural resources (Hufschmidt et al., 1982). Each of the surrogate methods has drawbacks, and at present there is no general agreement among economists concerning the most appropriate technique.

298. In the absence of accepted methods for estimating non-market values, researchers in this field have generally fallen back on using actual costs of maintenance and restoration of cultural property. It is recognized that this approach risks undervaluing the true costs of damage, since it does not include the non-market values, but it does give some indication of the magnitude of the effort to protect cultural property. An estimate in the Federal Republic of Germany of damages to outdoor sculpture and stained glass was in the range of DM 25-35 million per year. A study of restoration costs in the state of Massachusetts (USA) came up with an annual cost of $5 million per year for historic buildings and $0.6 million for outdoor sculpture (Rae, 1984). In the Netherlands, annual restoration costs were estimated at 15-30 million guilders, in addition to 0.25 million guilders for retuning carillon bells (Feenstra, 1984). On a common basis, in United States dollars, these figures are $0.14-0.19 per capita, $0.30-$0.60 per capita and $0.90 per capita for the Federal Republic of Germany, the Netherlands and United States of America, respectively.

299. While these numbers indicate the magnitude of the problem, they should not be regarded as comprehensive in scope or even comparable one with the other. The United States study emphasized maintenance costs, such as mortar repointing, but also includes major restoration projects, using a discount factor of 10 per cent. The Federal Republic of Germany study did not annualize capital costs for major restoration projects such as Cologne Cathedral, but listed them separately. The Dutch estimate was based on an annual maintenance cost of buildings, primarily repainting, repointing and roof repair. The backlog of major restoration work in the Netherlands is estimated at 200 million guilders.

300. Besides the lack of a consistent framework for comparing restoration costs, for cultural property damages there are some basic questions about using this type of data to estimate benefits. In most cases, the restoration work is not undertaken solely for the purpose of repairing damage from air pollution. Structural repairs, elimination of biological growths or correction of earlier restoration efforts are also reasons for undertaking restoration. Generally, all this is done simultaneously since the greatest single cost in such activities is putting up the scaffolding. It is often difficult to determine what portion of the total restoration cost should be assigned to air pollution damage, especially when a detailed breakdown of the costs is not available. It would be reasonable to blame air pollution for some share of the costs of faëade cleaning, mortar repointing, corrosion repair and repainting, but the exact proportion would depend upon many site-specific variables, as discussed under service life factors above. However, it would not be reasonable to include the costs of removal of lichens and mosses, the bracing of structural members cracked by settling, or the replacement of anachronistic materials.

301. Even if it is possible to assign costs to air pollution damage, it may not be appropriate to project future benefits using this data. As pointed out by Feenstra, the restoration work now going on represents a backlog that has accumulated over many years. Much of the air pollution damage now being repaired occurred in the past when pollution levels may have been much higher. Consequently, one cannot simply assume a linear relationship between costs of restoration and current air pollution levels. Instead, the costs are a function of the cumulative exposure to air pollution over a period in the lifetime of the monument, in other words, the air pollution dose. In order to calculate the dose, it is necessary to estimate historical levels of air pollution, which then becomes an exercise in pollution archaeology.

302. In summary, the economic valuation of air pollution damage to historic buildings and cultural monuments is an essential step in a cost-benefit analysis, but much work remains to be done, both in the development of analytical methods consistent with economic theory and in data collection. Current developments in decision - analysis which allows probability distributions to be assigned to estimates of various forms of environmental damage, offer an analytical tool to accommodate differences in scientific opinion with regard to such damage. Such analysis also enables the decision maker(s) to analyse the different set of values implicit in alternative policy options.

B. MODERN BUILDINGS/STRUCTURES

303. The benefits associated with reduction of damages to ordinary architecture can be calculated using standard life-cycle cost methodology (Ruegg et al. 1978). The reduction of air pollution should slow the rate of damage to a given material and thus extend its service life. The capital cost of replacement is thus spread out over a longer time period, thereby lowering annualized costs or present value. The operating or maintenance costs may be reduced as well, which would also be reflected in lower annualized costs or present value. Thus the economic benefit of reduced air pollution damage could be defined in terms of the difference in present value or annualized costs between the original situation and the level after controls have been put into effect (Weber, 1985).

304. This simple life-cycle cost model ignores some complications. As noted in Section II (paragraph 287 ff.), the service life itself may be determined by economic rather than independent physical factors. The conventional life-cycle model usually assumes that as a component reaches the end of its service life, it will be replaced by an identical new component. However, this like-for-like replacement may not occur, if the property owner chooses a more air-pollution resistant alternative (McCarthy et al.,1984). If the costs of the alternatives are known, it may be possible to incorporate a module into the life-cycle cost model that simulates the possible outcomes on a probabilistic basis (Holden, 1984). It should also be noted that in many cases the builder, who chooses the materials of construction, may not be the long-term owner. The builder's objective would be to minimize construction costs, even though this might increase operating and maintenance expenses over the life of the building. The benefits are also sensitive to the discount factors used in the life-cycle cost calculations. A high discount rate will reduce the significance of a given change in service life. Finally, a rigorous economic analysis should also take into account the fact that prices are a function of supply and demand and that economic benefits cannot simply be calculated as changes in physical damage X constant prices. For example, reductions in damage lower the demand for building repair services. The cost of these services would then decrease, thereby affecting calculated benefits if a theoretically correct model could be computed in terms of consumer and producer surpluses (Horst et al., 1984). However, little is known about price elasticities in this situation, and the data requirements to construct the necessary supply and demand curves are large. It may be that the changes in demand for building repair services as a result of reduced air pollution may be too small to affect price levels significantly.

305. A life-cycle cost approach has the virtue of putting the air pollution problem into the context of the specific economic decisions facing the individual building owner.

IV. CONCLUSIONS

306. **Service Life Factors** - Service life changes as a function of air pollution are estimated in one-dimensional terms, but in reality are determined by two-dimensional surface considerations. Little data are available.

307. **Materials inventory** - The materials inventory appears to be dominated by those materials used in residential construction, namely painted surfaces and brick. The ordinary materials can be characterized by a statistical model of materials distribution. However, specialized inventories may be required for cultural property, high-rise business districts and transportation networks.

308. **Economic valuation** - The economic valuation of materials effects is still evolving. For ordinary architecture, there is a need for a standard methodology for life-cycle costing. There are some problems to be solved regarding the valuation of damage to cultural property, particularly the non-market values.

REFERENCES

1. ASTM (American Society for Testing of Materials) "Standard Methods of Evaluating Degree of Rusting on Painted Surfaces" Annual Book of ASTM Standards, Part 27, (D-610, 1983).

2. ASTM (American Society for Testing of Materials) "Index to Selection and Use of Testing Procedures for Architectural Paints", Annual Book of ASTM Standards, Part 27, (D-2833, 1984).

3. R. Barnes, G. Parkinson, and A. Smith "The Costs and Benefits of Sulphur Oxide Control" Journal of the Air Pollution Control Association, Vol. 33, No. 8, (pp. 737-741, 1983).

4. ASTM (American Society for Testing of Materials) "Index to Selection and Use of Testing Procedures for Architectural Paints", Annual Book of ASTM Standards, Part 27, (D-2833, 1984).

5. R. Bastian, "Urban House Types as a Research Focus in Historical Geography", Environmental Review, Vol. 4, No. 2, (pp. 27-34, 1980).

6. ECOTECH Ltd. Identification and Assessment of Materials Damage to Buildings and Historical Monuments by Air Pollution, 2nd Progress Report, (Manchester, United Kingdom, 1985).

7. EPRI (Electric Power Research Institute) Air Pollution Damage to Man-made Materials: Physical and Economic Estimates, EPRI EA-2837 (Palo Alto, California, USA, p. 12, 1983).

8. J. Feenstra, Cultural Property and Air Pollution: Damage to Monuments, Art Objects, archives and buildings due to Air Pollution, (Ministry of Housing, Physical Planning and Environment, The Netherlands, pp. 122-123, 1984).

9. F. Haynie, J. Spence and J. Upham, Effects of Gaseous Pollutants on Materials - A Chamber Study, (United States Environmental Protection Agency, Research Triangle Park, North Carolina, 1976).

10. R. Holden, Decisions Confronting Owners of Older Buildings: An Economic Choice Model, (National Trust for Historic Preservation, Washington, DC, 1984).

11. R. Horst, E. Manuel and J. Bentley, Economic Benefits of Reduced Acidic Deposition on Common Building Materials: Methods Assessment, (Mathtech, Inc., Princeton, New Jersey, USA, 1984).

12. M. Hufschmidt, and others, Environment, Natural Systems and Development: An Economic Valuation Guide, (Johns Hopkins University Press, 1983) Chapter 6.

13. Leman Group Inc. Acid Rain Impact on the Urban Environment: Phase I, Methodology, (Toronto, Ontario, Canada, 1985).

14. F. Lipfert, and M. Torpey, "Methods for Materials Inventorying in High-rise Central Business Districts", (Brookhaven National Laboratory, Upton, New York, USA, 1984).

15. J. Martin, and M. McKnight, "The Prediction of Service Life of Coatings on Steel: Part 1 - Procedure for Quantitative Evaluation of Coating Defects" Journal of Coatings Technology, (July, 1985).

16. J. Martin, and M. McKnight, "The Prediction of the Service Life of Coatings on Steel: Part II - Quantitative Prediction of the Service Life of a Coating System", Journal of Coatings Technology, (July, 1985) op.cit. (15)

17. F. Matero, and J. Teutonico, "The Use of Architectural Sandstone in New York City in the 19th Century", Bulletin of the Association for Preservation Technology, (Vol. XIV, No. 2, pp. 11-17, 1982).

18. E. McCarthy, and others, Damage Cost Models for Pollution Effects on Materials, PB 84-140-342, (National Technical Information Service, Springfield, Virginia, USA, 1984).

19. C. Merry, and P. LaPotin, Regression Models for Predicting Building Material Distribution in Four North-eastern Cities, (Regions Research and Engineering Laboratory, Hanover, New Hampshire, 1985).

20. K. Novak, and others, National Data Bases for Residential Exterior Building Materials, (Brookhaven National Laboratory, Upton, New York, USA, 1985).

21. D. Rae, "Economic Valuation of Cultural Materials: Some Preliminary Findings" 77th Annual Meeting of the Air Pollution Control Association, Paper 84-83.5, (Pittsburgh, 1984).

22. D. Repellin, "The Organization of Historic Building Documentation in France", Proceedings of the International Seminar on Architectural and Engineering Documentation, (US/ICOMOS, 1983).

23.   G. Rosenfield, "Spatial Sampling Design for Building Materials Inventory for Use with an Acid Rain Damage Survey", in R. Schmidt and H. Smolin, eds., The Changing Role of Computers in Public Agencies, (The Urban and Regional Information Systems Association, 1984).

24.   R. Ruegg, and others, "Life-Cycle Costing: A Guide for Selecting Energy Conservation Projects for Public Buildings", NBS Building Science Series 113, (United States National Bureau of Standards, Gaithersburg, Maryland, pp. 1-15, 1978).

25.   S. Sherwood, "The National Park Service Research Program on the Effects of Air Pollution on Cultural Properties", 77th Annual Meeting of the Air Pollution Control Association, Paper No.83-1, (1984).

26.   J. Strenio, and others, Asbestos in Buildings: A National Survey of Asbestos-Containing Friable Materials (United States Environmental Protection Agency, Washington, DC, pp. 4-1 to 5-3, 1984).

27.   G. Taylor, Managerial and Engineering Economy, (Van Nostrand, New York, pp. 163-187, 1968).

28.   P. Thurber, "Controversies in Historic Preservation: Understanding the Preservation Movement Today", (National Trust for Historic Preservation, Washington, DC, 1985).

29.   M. Torpey, and F. Lipfert, A Materials Inventory of Infrastructure Steel: A Methodology, (Brookhaven National Laboratory, Upton, New York, 1984).

30.   TRC Environmental Consultants Inc., Assessment of Materials Damage and Soiling from Air Pollution in the South Coast Air Basin, (TRC, Hertfort, Connecticut, 1985).

31.   J. Weber, "Natural and Artificial Weathering of Austrian Building Stones due to Air Pollution", Fifth International Congress on the Deterioration and Conservation of Building Stones, (Lausanne, Switzerland, 1985).

32.   M. Wills, Documenting Historic Buildings in Bavaria: 100 years of Experience", Proceedings of the International Seminar on Architectural and Engineering Documentation, (US/ICOMOS, 1983).

Part TWO

Technologies for Controlling Sulphur and Nitrogen Emissions

Chapter 5

CONTROL OF SULPHUR DIOXIDE EMISSIONS FROM INDUSTRIAL PROCESSES

309. Although the largest contributors to anthropogenic SO_2 are utility and industrial boilers, significant amounts of SO_2 emissions can also originate in industrial processes such as non-ferrous metal smelting, petroleum refining, cement manufacturing, iron and steel manufacturing, and pulp and paper production.

310. For the purpose of this report, emissions from industrial processes are understood as emissions resulting from the industrial treatment, processing or conversion of raw materials or intermediate products, excluding emissions from combustion of fossil fuels for energy generation and from accidents or leaks. Technologies for controlling sulphur emissions from combustion of fossil fuels have already been described and assessed in detail elsewhere (1) (2). Emissions resulting from accidents or leaks are rather difficult to consider in a systematic way because numerous factors influence the choice of control options.

NON-FERROUS METAL SMELTING (3)

A. PROCESS DESCRIPTIONS

311. Efforts to control air pollutant emissions from non-ferrous metal production processes have traditionally been focused upon the problem of SO_2 control. The SO_2 emissions from smelters fall roughly into two categories, strong and weak (i.e. greater or less than 4 per cent SO_2 respectively). This terminology arises from the application of a sulphuric acid plant to control smelter SO_2 emissions. Such plants require a minimum SO_2 gas-strength in order to be economically feasible. Consequently, an off-gas stream having a sufficiently high SO_2 concentration for a sulphuric acid plant is a strong SO_2 off-gas. Those SO_2 off-gas streams having concentrations less than the minimum for sulphuric acid production are called weak SO_2 off-gas streams.

312. The type of ore and its metal values determine the choice of metallurgical process used for its extraction. Usually, there are present additional metallic elements such as gold, silver, etc. which are also to be recovered in subsequent processing steps. Several processes have been developed for copper and zinc and a few for lead and nickel. The processes fall into two basic classifications: pyrometallurgical and hydrometallurgical. Hydrometallurgical processes are based upon the solution of the metal values in an aqueous medium followed by electrochemical winning of the metal values from the solution. Hydrometallurgical processes are used principally for oxide ores or materials converted to oxides. With the exception of zinc production, hydrometallurgical processes are not applied as broadly for large-scale systems as are pyrometallurgical processes. In pyrometallurgical processing, the metal values are extracted by slagging and volatilizing of the undesirable components. The sulphur contained in the ore concentrate is a fuel which furnishes heat for the pyrometallurgical processes. More precisely, it is volatilized and oxidized, resulting in substantial sulphur dioxide emissions.

(a) Copper processing

313. Copper-smelting plants are characterized by the use of ore as process material. Ores can be classified into sulphide, oxide or sulphide-oxide. Copper sulphides account for 80 per cent of the world's copper ore reserves.

314. Sulphide copper ore concentrates and the richer oxidic copper ores are treated primarily by the pyrometallurgical process. Hydrometallurgical techniques for extracting copper are applied mainly to low-grade oxide

copper ores, sulphide-oxide ores, and some complex ores. Pyrometallurgical processes are a main source of SO_2 emissions.

315. Copper smelting consists of either two or three distinct pyrometallurgical processing steps: (a) roasting, (b) smelting, and (c) converting. Sulphur in excess of the amount needed to ensure formation of a copper sulphide matte in the furnace can be eliminated by roasting. The calcine produced by the roaster, or the green concentrate if roasting is not used, is charged to the smelting furnace where the charge is forced through complex reactions involving melting, slagging and volatilization of impurities to form a copper matte.

316. Copper matte is a mixture of molten iron sulphide and copper sulphide (Cu_2S). The relative proportions of the iron sulphide and copper sulphide vary over a wide range. The iron sulphide is preferentially oxidized to iron oxide and combines with a silica slagging agent to form a liquid iron silicate slag that is immiscible with the sulphide matte phase. The slag phase floats on top of the matte layer. As excess slag is formed, it is removed and discarded. The matte is periodically tapped and is further processed in a convertor in which air is blown through the smolten metal to remove iron and the other impurities and form blister copper. About 1-2 per cent of the sulphur entering a smelter is lost in slags, 3-4 per cent is released as fugitive emissions, and the remainder is contained as SO_2 in the gases from roasters, smelting furnaces, and converters. A total of about two tonnes of SO_2 is generated for each tonne of copper produced.

317. Although the three steps of copper smelting have the same functions in all smelters, there are significant differences in the equipment used, the operating conditions and the intermediate products and emissions produced.

318. In copper smelting, the major sources of weak SO_2 off-gas are the reverberatory furnace and multihearth roaster, followed by the fugitive emissions occurring from (a) the converter operation, (b) matte tapping, (c) slag tapping, and (d) ladle transfer.

(b) Nickel processing

319. The metallurgy of nickel processing is similar to copper and the same type of processing equipment is used. The sources of sulphur dioxide emissions are the same. However, significantly more SO_2 is generated per tonne of metal produced than in the case of copper. Approximately 10 tonnes of SO_2 is generated overall per tonne of nickel produced. This is owing to the large amounts of pyrrholite

associated with nickel ore bodies. A substantial portion of the pyrrholite is separated in the mineral beneficiation stage. This may be subsequently processed in order to recover residual nickel, sulphuric acid and iron ore.

(c) Lead processing

320. Metallic lead is produced from the concentrates by a sequence of processes, consisting of bullion production and bullion refining. The most important bullion production method with regard to the tonnage produced and frequency of application is sintering and blast furnace smelting. A number of processes have been developed, in recent years, which eliminates the need for sintering and permits direct smelting. They thereby have environmental advantages.

321. In most cases, the first processing step in a lead smelter is sintering. Sintering is performed to remove sulphur and to produce a material suitable for charging to the shaft furnace. About 85 per cent of the sulphur is removed by oxidation and the feed is fused into a porous clinker. The product from sintering is then processed in a shaft furnace to produce lead bullion, an impure metal. Sintering is the only part of the lead smelting process than can emit large amounts of SO_2. Fugitive emissions of SO_2 also occur at the discharge end of the sinter machine, where the discharged material is broken into pieces. Low concentrations of SO_2 may occur in shaft furnace off-gas. Processes that purify lead bullion emit no SO_2, although they do emit other pollutants.

322. In lead smelting, about 85 per cent of the sulphur in the concentrate is liberated as SO_2 in the sintering step. Of the remaining 15 per cent sulphur in the concentrate, around half is eliminated as SO_2 in the off-gases from the subsequent processing steps and half in the solid products. In the sintering process, most of the sulphur is eliminated at the front end of the sinter machine. The strong gas stream is usually collected and controlled by a metallurgical acid plant. Some smelters have installed a recirculation system for the weak stream which allows its combination with the strong stream.

(d) Zinc processing

323. The producton of zinc metal from concentrates can be separated into two basic methods, electrolytic refining and pyrometallurgical smelting. Electrolysis currently accounts for more than one half of the world's zinc production. In either method, the zinc concentrate is first roasted pyrometallurgically to drive off unwanted sulphur, lead, and other

constituents that could interfere with subsequent processing.

324. During roasting, at least 95 per cent of the sulphur is converted into strong SO_2 and most of the remaining sulphur is converted into sulphates. The product of roasting, called calcine, is an impure zinc oxide, which usually contains less than 0.3 per cent sulphide sulphur. Roasting is the only significant source of SO_2 emission in a zinc smelter. The electrolytical process produces no SO_2 emissions. Pyrometallurgical production may create low concentrations of SO_2 (< 0.1 per cent) in off-gas from sintering machines (Figure 1).

B. CONTROL TECHNOLOGY

325. As pointed out above, off-gases from non-ferrous smelters basically fall into two categories: those with high SO_2 concentrations (defined as greater than 4 per cent SO_2) and those with low SO_2 concentrations. Strong gas streams are controllable using add-on technologies such as acid plants and liquid SO_2 plants. These processes are considered as proven and, in most cases, economically feasible control options. Treatment of weak gas streams constitutes a more difficult and costly problem. The available options are:

- Use of the add-on technology of flue-gas desulphurization (FGD);

- Modification of the furnaces to produce a strong gas stream through measures such as oxygen enrichment; and

- Replacement of sources giving rise to the weak SO_2 stream with alternative, modern technology producing strong SO_2 streams controllable by acid plants.

(a) Use of add-on technology

326. Flue gas desulphurization (FGD) is practised by a number of smelters worldwide. The FGD systems employed are unique to each smelter. This is a result of the particular circumstances of each application in terms of economies for raw materials and by-product markets rather than technical suitability of the process. The flue gas desulphurization processes applied can be classified as recovery or non-recovery processes. The LINDE process using an organic solvent, the magnesium oxide scrubbing process and the sodium sulphite (WELLMAN-LORD) scrubbing process should only be mentioned as examples of industrially applied recovery processes.

(b) Process modification

327. Upgrading of existing furnace operations to produce stronger SO_2 streams can be an effective approach to SO_2 control when coupled with acid plants. Alternative pyrometallurgical processes are of interest because they provide a strong SO_2 gas stream controllable by a conventional acid plant, reduction in energy consumption, lower gas stream volumes, and less operating costs. For instance, direct smelting of lead avoiding the need for sintering offers the advantage of: reducing the number and volume of off-gas streams to be cleaned; eliminating the low strength SO_2 stream from the blast furnace for which SO_2 recovery is usually impractical; and reducing the number of fugitive emissions requiring ventilation.

(c) Process substitution

328. A recent technology for the reduction of sulphur emissions is flash smelting. The flash smelting process for sulphidic concentrates is designed to replace the conventional method of separate roasting and smelting, by combining roasting, smelting and also, partly, converting processes into a single suspension smelting process. The dried, fine-grained concentrate is fed into the flash smelting furnace together with oxygen or oxygen-enriched process-air to form a suspension. The heat generated by the rapid exothermic oxidation reaction in the suspension simultaneously smelts the charge. Process modifications exist for smelting of copper and nickel concentrates, and now lead concentrates as well. The high SO_2 concentration in the effluent gas is suitable for the production of elemental sulphur, liquid SO_2 or sulphuric acid.

329. A large number of alternative approaches to achieve reductions in SO_2 emissions based upon various combinations of process and control technologies are technically possible. In any approach, it is of paramount importance to consider the unique nature of each smelter. This uniqueness factor is determined by the nature of the ore concentrates and metallurgy required to treat successfully these concentrates. These aspects govern the selection of a metallurgical process for metal winning, and, in turn, the degree of sulphur containment. Each smelter requires an individual technical and economic assessment of feasibility.

FIGURE 1

Generalized flow diagrams for selected pyrometallurgical processes

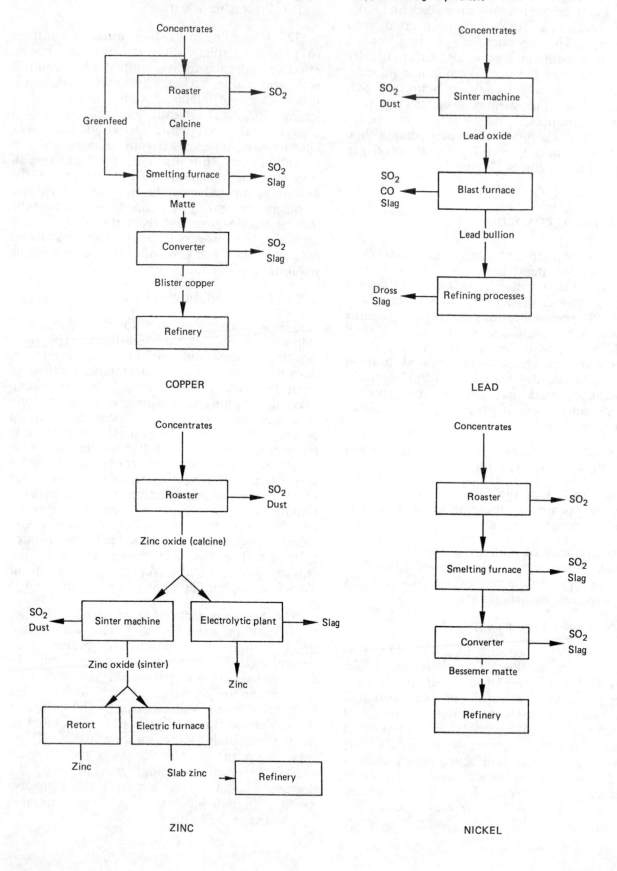

330. Alternative hydrometallurgical processes eliminate the generation of uncontrollable SO_2 streams. The capital costs of hydrometallurgical processes are competitive with pyrometallurgical processes. The operating costs may be higher depending on the cost of electric power. However, electrolytic cell technology advances are being made in battery and fuel cells which may have important implications for hydrometallurgy.

331. The application of most control technologies, excepting acid plants, tends to be limited by economic aspects linked to the market for the by-product produced, the capital and operating costs, or lack of extensive experience with certain technologies, especially those for control of weak SO_2 streams.

C. COSTS OF SULPHUR EMISSION CONTROL (4)

332. The cost of abating sulphur dioxide emissions varies according to the control technology used. The cost of implementation of each technology at different smelters is also widely variable depending on the smelting process used, the amount of smelter retrofit required, availability of markets for by-products, off-gas flow rates, SO_2 content, etc. The following general points should be noted:

(a) According to North American estimates, sulphur dioxide abatement using contact sulphuric acid plants provides the lowest cost per tonne of sulphur removed ($US 77 to 121) given an adequate SO_2 concentration. This estimate assumes that the acid can be sold at a price sufficient to cover the costs of transportation and marketing. In some instances, depending upon smelter location and competition for markets, the sales revenue may be insufficient to cover distribution costs which would then have to be borne by the smelter, increasing the cost of SO_2 fixation. In some instances, neutralization may be required which would increase costs further;

(b) Process modifications to provide gases with higher SO_2 concentrations and thereby enable fixation in acid plants offer a viable alternative for SO_2 control. However, no cost estimates are available for such modifications;

(c) Treatment of dilute gases using FGD processes have the highest operating costs per tonne of sulphur removed. The cost for the regenerative options are lower than in the case of non-regenerative options as a result of a reagent recovery and consequent cost savings; and

(d) The cost per tonne of sulphur removed for the process replacement options which enable gases to be treated in sulphuric acid plants is estimated to be considerable ($US 267). The major cost component is capital amortization.

II. PETROLEUM REFINING

333. The petroleum refining industry employs a wide variety of processes to convert oil into more than 2,500 refined products. A refinery's processing flow scheme is largely determined by the composition of the crude oil feed stock and the chosen array of petroleum products.

A. PROCESS DESCRIPTIONS (5)(6)(7)(8)

(a) Catalytic cracking

334. Catalytic cracking, using heat, pressure and catalyst, converts heavy oils into lighter products with product distributions favouring the more valuable gasoline and distillate blending components. Feedstocks are usually gas oils from atmospheric distillation, vacuum distillation, coking, and de-asphalting processes. These feedstocks typically have a boiling range of 340 to 540°C. All of the catalytic cracking processes in use today can be classified as either fluidized-bed or moving-bed units.

(i) Fluidized-bed catalytic cracking (FCC)

335. The FCC process uses a catalyst in the form of very fine particles that act as a fluid when aerated with vapour. Fresh feed is preheated in a process heater and introduced into the bottom of a vertical transfer line or riser with hot regenerated catalyst. The hot catalyst vaporizes the feed bringing both to the desired reaction temperature of 470 to 525°C. The high activity of modern catalysts causes most of the cracking reactions to take place in the riser as the catalyst and oil mixture flows upward into the reactor. The hydrocarbon vapours are separated from the catalyst particles by cyclones in the reactor. The reaction products are sent to a fractionator for separation.

336. The spent catalyst falls to the bottom of the reactor and is steam stripped as it exits the

reactor bottom to remove absorbed hydrocarbons. The spent catalyst is then conveyed to a regenerator. In the regenerator, coke deposited on the catalyst as a result of the cracking reactions is burned off in a controlled combustion process with preheated air. Regenerator temperature is usually 590 to 675°C. The catalyst is then recycled to be mixed with fresh hydrocarbon feed.

(ii) *Moving-bed catalytic cracking*

337. In the moving-bed catalytic cracking process, catalyst beads (approx. 0.5 cm) flow by gravity into the top of the reactor where they contact a mixed-phase hydrocarbon feed. Cracking reactions take place as the catalyst and hydrocarbons move concurrently downward through the reactor to a zone where the catalyst is separated from the vapours. The gaseous reaction products flow out of the reactor to the fractionation section of the unit. The catalyst is steam-stripped to remove any absorbed hydrocarbons. It then falls into the regenerator where coke is burned from the catalyst with air. The regenerated catalyst is separated from the flue gases and recycled to be mixed with fresh hydrocarbon feed. The operating temperatures of the reactor and regenerator in the TCC process are comparable to those in the FCC process.

338. SO₂ emissions from catalytic cracking processes are (a) combustion products from process heaters and (b) flue gas from catalyst regeneration. Many refiners use a CO-burning waste-heat boiler to recover the energy contained in the flue gas. In addition to energy recovery, this procedure reduces CO emissions to low levels. Other combustible contaminants may be removed, depending on the operating temperature of the boiler. Increased emissions of thermal NO$_x$ may, however, result with higher boiler temperatures, and sulphur in the auxiliary fuel will be converted to SO$_x$. In some installations, sulphur oxides are removed by passing the regenerator flue gases through a water or caustic scrubber.

(b) Thermal cracking

339. Thermal cracking processes include visbreaking and coking, which break heavy oil molecules by exposing them to high temperatures.

(i) *Visbreaking*

340. Topped crude or vacuum residuals are heated and thermally cracked at 455° to 480°C, 3.5 to 17.6 kg/cm² in the visbreaker furnace to reduce viscosity or pour point of the charge. The cracked products are quenched with gas oil and flashed into a fractionator. The vapour overhead from the fractionator is separated into light distillate products. A heavy distillate recovered from the fractionator liquid can be used as a fuel oil blending component or used as catalytic cracking feed.

(ii) *Coking*

341. Coking is a thermal cracking process used to convert low value residual fuel oil into higher value gas oil and petroleum coke. Vacuum residuals and thermal tars are cracked in the coking process at high temperature and low pressure. Products are petroleum coke, gas oils, and lighter petroleum stocks. Delayed coking is the most widely used process today, but fluid coking is expected to become an important process in the future.

342. In the delayed coking process, heated charge stock is fed into the bottom section of a fractionator where light ends are stripped from the feed. The stripped feed is then combined with recycle products from the coke drum and rapidly heated in the coking heater to a temperature of 480 to 590°C. Steam injection is used to control the residence time in the heater. The vapour-liquid feed leaves the heater, passing to a coke drum where, with controlled residence time, pressure (1.8 to 2.1 kg/cm²), and temperature (400°C), it is cracked to form coke and vapours. Vapours from the drum return to the fractionator where the thermal cracking products are recovered.

343. In the fluid coking process, residual oil feeds are injected into the reactor where they are thermally cracked, yielding coke and a wide range of vapour products. Vapours leave the reactor and are quenched in a scrubber where entrained coke fines are removed. The vapours are then fractionated. Coke from the reactor enters a heater and is devolatilized. The volatiles from the heater are treated for fines and sulphur removal to yield a particulate-free, low-sulphur fuel gas. The devolatilized coke is circulated from the heater to a gasifier where 95 per cent of the reactor coke is gasified at high temperature with steam and air or oxygen. The gaseous products and coke from the gasifier are returned to the heater to supply heat for the devolatilization. These gases exit the heater with the heater volatiles through the same fines and sulphur removal processes.

344. From available literature, it is unclear what emissions are released and where they are released. Air emissions from thermal cracking processes include coke dust from decoking operations, combustion gases from the

visbreaking and coking process heaters, and fugitive emissions. Fugitive emissions from miscellaneous leaks are significant because of the high temperatures involved, and are dependent upon equipment type and configuration, operating conditions, and general maintenance practices. Particulate emissions from delayed coking operations are potentially very significant. These emissions are associated with removing the coke from the coke drum and subsequent handling and storage operations. Hydrocarbon emissions are also associated with cooling and venting the coke drum prior to coke removal. However, comprehensive data for delayed coking emissions have not appeared in available literature.

345. The flue gas from the burner bed normally contains significant quantities of CO and has a heating value of around 1,622 kJ/Nm³. Hence, this stream is usually discharged through cyclones to a waste heat recovery device such as a CO boiler or process heater.

(c) Utilities plant

346. The utilities plant supplies the steam necessary for the refinery. Although steam can be used to produce electricity by throttling it through a turbine, it is primarily used for heating and separating hydrocarbon streams. When used for heating, the steam usually heats the petroleum indirectly in heat exchangers then returns to the boiler. In direct contact operations, the steam can serve as a stripping medium or a process fluid. Steam may also be used in vacuum ejectors to produce a vacuum. Emissions from boilers and applicable emission control technology are well known and will not be repeated here.

(d) Process heaters

347. Process heaters (furnaces) are used extensively in refineries to supply the heat necessary to raise the temperature of feed materials to reaction or distillation level. They are designed to raise petroleum fluid temperatures to a maximum of about 510°C. The fuel burned may be refinery gas, natural gas, residual fuel oils, or combinations, depending on economics, operating conditions and emission requirements. Process heaters may also use carbon monoxide rich regenerator flue gas as fuel.

348. Particulates, sulphur and nitrogen oxides, hydrocarbons, and carbon monoxide are all emitted from process heaters. The quantity of these emissions is a function of the type of fuel burned, the nature of the contaminants in the fuel, and the heat duty of the furnace. Sulphur oxides can be controlled by fuel desulphurization or flue gas treatment. For instance, the sodium sulphite scrubbing process (WELLMAN-LORD process) has been applied for the desulphurization of flue gases linked with a Claus unit for the recovery of elemental sulphur.

(e) Sulphur recovery plant

349. A sulphur recovery plant converts hydrogen sulphide (H₂S) separated from refinery gas streams into the more disposable by-product, elemental sulphur. Most of the elemental sulphur produced from H₂S is made by the modified Claus process. A simplified flow diagram of this process is shown in Figure 2. The process consists of the multistage catalytic oxidation of hydrogen sulphide according to the following overall reaction:

$$2H_2S + O_2 ---> 2S + 2H_2O \ (1)$$

In the first step, one third of the H₂S is reacted with air in a furnace and combusted to SO₂ according to reaction (2):

$$H_2S + 1\ 1/2\ O_2 ---> SO_2 + H_2O \ (2)$$

The heat of the reaction is recovered in a waste heat boiler or sulphur condenser.

350. For gas streams with low concentrations of H₂ (20 per cent to 60 per cent), approximately one third of the gas stream is fed to the furnace and the H₂S is nearly completely combusted to SO₂, while the remainder of the gas is bypassed around the furnace. This is the "split stream" configuration. For gas streams with higher H₂S concentrations, the entire gas stream is fed to the furnace with just enough air to combust one third of the H₂S to SO₂. This is the "partial combustion" configuration. In this configuration as much as 50 to 60 per cent conversion of the hydrogen sulphide into elemental sulphur takes place in the initial reaction chamber by reaction (1). In extremely low concentrations of H₂S (less than 25 per cent to 30 per cent) a Claus process variation known as "sulphur recycle" may be used, where product sulphur is recycled to the furnace and burned, raising the effective sulphur level where flame stability may be maintained in the furnaces.

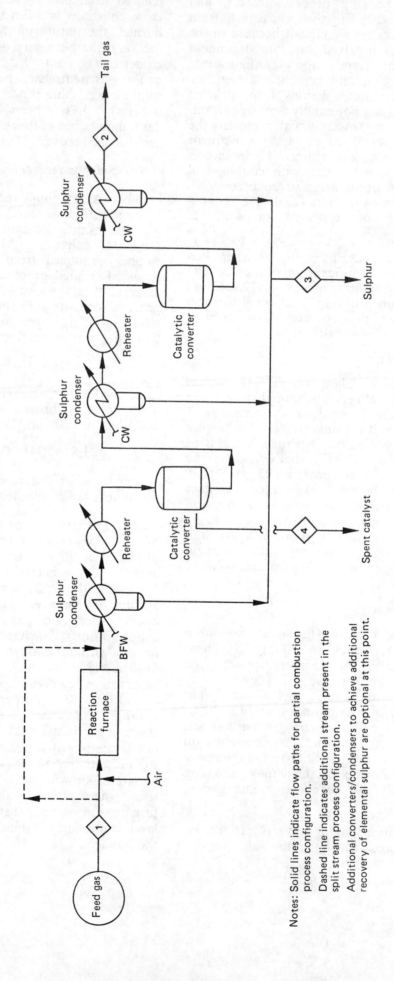

FIGURE 2

Typical flow diagram - Claus Process sulphur recovery

Notes: Solid lines indicate flow paths for partial combustion process configuration.

Dashed line indicates additional stream present in the split stream process configuration.

Additional converters/condensers to achieve additional recovery of elemental sulphur are optional at this point.

351. After the reaction furnace phase, the gases are cooled to remove elemental sulphur and then reheated. The remaining H_2S in the gas stream is then reacted with the SO_2 over a bauxite catalyst at 260 to 316°C to produce elemental sulphur according to reaction 3:

$$2H_2S + SO_2 \text{ --->} 3S + 2H_2O \text{ (3)}$$

Because this is a reversible reaction, equilibrium requirements limit the conversion. Lower temperatures favour elemental sulphur formation, but at too low a temperature elemental sulphur fouls the catalyst. Because the reaction is exothermic, the conversion attainable in one stage is limited. Therefore, two or more stages are used in series, with interstage cooling to remove the heat of reaction and to condense the sulphur.

352. Carbonyl sulphide (COS) and carbon disulphide (CS_2) are formed in the reaction furnace in the presence of carbon dioxide and hydrocarbons:

$$CO_2 + H_2S \text{ --->} H_2O + COS \text{ (4)}$$
$$COS + H_2S \text{ --->} H_2O + CS_2 \text{ (5)}$$
$$CH_4 + 4S \text{ --->} CS_2 + 2H_2S \text{ (6)}$$

About 0.25 to 2.5 per cent of the sulphur fed may be lost in this way. Additional sulphur may be lost as vapour, mist or droplets.

353. Tail gas from a Claus sulphur recovery unit contains a variety of pollutants, including sulphur dioxide. If no other controls are used, the tail gas is incinerated, so that the emissions consist mostly of sulphur dioxide.

354. Sulphur dioxide emissions (along with H_2S and sulphur vapour) depend directly on the sulphur recovery efficiency of the Claus plant. This efficiency is dependent upon many factors, including the following:

- Number of catalytic conversion stages

- Inlet feed stream composition

- Operating temperatures and catalyst maintenance

- Maintenance of the proper stoichiometric ratio of H_2S/SO_2

- Operating capacity factor.

355. Recovery efficiency increases with the number of catalytic stages used. For example, for a Claus plant fed with 90 per cent H_2S, sulphur recovery is approximately 85 per cent for one catalytic stage and 95 per cent for two or three stages.

356. Recovery efficiency also depends on the inlet feed stream composition. Sulphur recovery increases with increasing H_2S concentration in the feed stream. For example, a plant having two or three catalytic stages would have a sulphur recovery efficiency of approximately 90 per cent when treating a 15 mole per cent H_2S feed stream, 93 per cent for a 50 mole per cent H_2S stream, and 95 per cent for a 90 mole per cent H_2S stream. Various contaminants in the feed gas reduce Claus sulphur recovery efficiency. Organic compounds in the feed require extra air for combustion. Added water and inert gas from burning these organics decrease sulphur concentrations and thus lower sulphur recovery rates. Higher molecular weight organics also reduce efficiencies because of soot formation on the catalyst. High concentrations of CO_2 in the feed gas reduce catalyst life.

357. The Claus reactions are exothermic which enhances sulphur recovery by removing heat and operating the reactors at as low a temperature as practicable without condensing sulphur on the catalyst. Recovery efficiency also depends on catalyst performance. One to two per cent loss in recovery efficiency over the period of catalyst life has been reported. Maintenance of the 2:1 stoichiometric ratio of H_2S to SO_2 is essential for efficient sulphur recovery. Deviation above or below this ratio results in a loss of efficiency. Operation of a Claus plant below capacity may also impair Claus efficiency somewhat.

358. Removal of sulphur compounds from Claus plant tail gas is possible by three general schemes:

- Extension of the Claus reaction to increase overall sulphur recovery,

- Conversion of sulphur gases to SO_2, followed by SO_2 removal technology,

- Conversion of sulphur gases to H_2S, followed by H_2S removal technology.

359. Processes in the first scheme remove additional sulphur compounds by carrying out the Claus reaction at lower temperatures to shift equilibrium of the Claus reactions towards formation of additional sulphur. The IFP-1, SBR/Selecox, Sulfreen and Amoco CBA processes use this technique to reduce the concentration of tail gas sulphur compounds to 1,500-2,500 ppm, thus increasing the sulphur recovery of the Claus plant to 99 per cent.

360. In the second class of processes, the tail gas is incinerated to convert all sulphur compounds to SO_2. The SO_2 is then recovered by one of several processes, such as the

Wellman-Lord. In the Wellman-Lord and certain other processes, the SO_2 absorbed from the tail gas is recycled to the Claus plant to recover additional sulphur. Processes in this class can reduce the concentration of sulphur compounds in the tail gas to 200-300 ppm, or less, for an overall sulphur recovery efficiency (including the Claus plant) of 99.9+ per cent.

361. The third method for removal of sulphur compounds from Claus tail gas involves converting the sulphur compounds to H_2S by mixing the tail gas with a reducing gas and passing it over a reducing catalyst. The H_2S is then removed, by the Stretford process (in the Beavon and Clean Air processes) or by an amine absorption system (SCOT process). The Beavon and Clean Air processes recover the H_2S as elemental sulphur, and the SCOT process produces a concentrated H_2S stream which is recycled to the Claus process. These processes reduce the concentration of sulphur compounds in the tail gas to 200-300 ppm or less and increase the overall recovery efficiency of the Claus plant to 99.9+ per cent.

362. In the United States, a New Source Performance Standard for Claus sulphur recovery plants in petroleum refineries was promulgated in March 1978. This standard limits emissions to 0.025 per cent by volume (250 ppm) of SO_2 on a dry basis and at zero per cent oxygen, or 0.001 per cent by volume of H_2S and 0.03 per cent by volume of H_2S, COS and CS_2 on a dry basis and at zero per cent oxygen.

B. COSTS OF CONTROLLING EMISSIONS

363. In response to a proposal from the Commission of the European Communities for a Council Directive (COM(83) 704 final) on the limitation of atmospheric pollutant emissions from large combustion plants of over 50 MW rated heat output, the oil companies' international study group for conservation of clean air and water - Europe (CONCAWE) set up a study group to examine potential control methods and estimate anticipated capital and operating costs to achieve the emission reductions for existing plant in oil refineries. (9)

364. The study was based on applying control measures to the "average" CONCAWE refinery. This average was derived from combustions and refinery fuel information, supplied by about 85 per cent of the participating oil refineries. According to the Directive, member countries would have full flexibility in how the overall reductions were to be achieved. However, for the purpose of the study it was assumed that the proposed percentage reductions would apply to all oil refineries, although this may not be the case in individual member countries.

365. Preliminary screening studies were carried out to indicate the most cost-effective means of controlling sulphur dioxide (SO_2), nitrogen oxides (NO_x) and particulates emissions from large combustion plants. Three options were considered for SO_2 control: flue gas desulphurization (FGD), fuel oil desulphurization and fuel oil gasification. FGD, despite very high associated capital and operating costs, is the least costly. A selection of five commercial FGD processes (one throwaway, two gypsum and two regenerable) was chosen to determine their economics in a refinery situation. De-NO_x processes and electrostatic precipitators (ESPs) were studied as being probably the most suitable means to achieve the proposed reductions in NO_x and particulates' emissions, respectively, from refineries.

366. The study assumed only single train units and a minimum additional retrofitting cost of 25 per cent on capital for all control facilities. The specific conditions in oil refineries required special consideration. This was because of the need to use the by-products of refinery processes as a fuel which may make it difficult to segregate fuel oil for burning. It was assumed that flue gas from fuel oil-firing can be largely segregated from that from fuel gas firing for FGD and ESP treatment. If this is not possible for safety, reliability or flexibility reasons, the cost given in the report will be substantially increased.

367. For the "average" CONCAWE refinery, the minimum retrofitting costs for reducing SO_2 emissions by the proposed 60 per cent using FGD were estimated at: capital investments $US 22-50 million and annual operating costs $US 10-16 million. The annual operating costs correspond to between $US 2,600 and 4,500 per metric tonne of sulphur removed.

368. It should be noted that waste disposal on a significant scale is required for many of the FGD processes and the disposal costs and space requirements can be considerable. Moreover experience with FGD within oil refineries is limited.

369. Retrofitting of controls to existing refineries would be difficult because of space limitations close to combustion plants and the multiplicity of stacks in some refineries. In those cases where space limitations are more severe than in the average CONCAWE refinery, costs could be considerably higher.

370. The above costs were considered accurate to an estimated range of -10/+40 per cent.

III. CEMENT MANUFACTURING

A. PROCESS DESCRIPTION

371. The more than 30 raw materials used to make cement may be divided into four basic components: lime (calcereous), silica (siliceous), alumina (argillaceous), and iron (ferriferous). Approximately 1,600 kg of dry raw materials are required to produce 1 tonne of cement. Approximately 35 per cent of the raw material weight is removed as carbon dioxide and water vapour. As shown in Figure 3 the raw materials undergo separate crushing after the quarrying operation, and when needed for processing, are proportioned, ground, and blended using either the wet or dry process.

372. In the dry process, the moisture content of the raw material is reduced to less than 1 per cent either before or during the grinding operation. The dried materials are then pulverized into a powder and fed directly into a rotary kiln. Usually, the kiln is a long, horizontal, steel cylinder with a refractory brick lining. The kilns are slightly inclined and rotate around the longitudinal axis. The pulverized raw materials are fed into the upper end and travel slowly to the lower end. The kilns are fired from the lower end so that the hot gases pass upward and through the raw material. Drying, decarbonating, and calcining are accomplished as the material travels through the heating kiln, finally burning to incipient fusion and forming the clinker. The clinker is cooled, mixed with about 5 per cent gypsum by weight, and ground to final product fineness. The cement is then stored for later packaging and shipment.

373. With the wet process, a slurry is made by adding water to the initial grinding operation. Proportioning may take place before or after the grinding step. After the materials are mixed, the excess water is removed and final adjustments are made to obtain a desired composition. This final homogeneous mixture is fed to the kilns as a slurry of 30 per cent to 40 per cent moisture or as a wet filtrate of about 20 per cent moisture. The burning, cooling, addition of gypsum, and storage are carried out as in the dry process.

B. EMISSIONS AND CONTROLS

374. Emissions of SO_x from cement kilns are predominantly SO_2. The sulphur content of the fuel and raw-material feed both contribute to the total SO_2 generated in the kiln. However, only a fraction of the SO_2 generated in the kiln is actually released to the atmosphere. The limestone in the raw-material feed produces calcium oxide and other alkaline oxides in the kiln. These alkaline materials react with much of the SO_2 to form calcium sulphate which is eventually incorporated in the solid clinker. Approximately 75 to 90 per cent of the SO_2 can be removed in this fashion. The alkaline content of the feed can influence the actual removal rates. The SO_2 emission rate can be significant when high sulphur fuels are used in conjunction with low alkali feed materials.

375. If added SO_2 control is needed, scrubbers can potentially achieve a 90 per cent reduction in SO_2 emissions. However, no scrubber has been installed in a commercially active cement plant. As a result, actual scrubber performance is uncertain and available cost estimates speculative. The best estimates of scrubber costs range from $1 to $2 per kg of SO_2 removed depending on kiln size and operating parameters.

376. Fabric filters can achieve a 50 per cent reduction in SO_2 emissions. Precipitators do not achieve as large a reduction in SO_2 as fabric filters. Fabric filter cost is $0.6 per kg of sulphur removed.

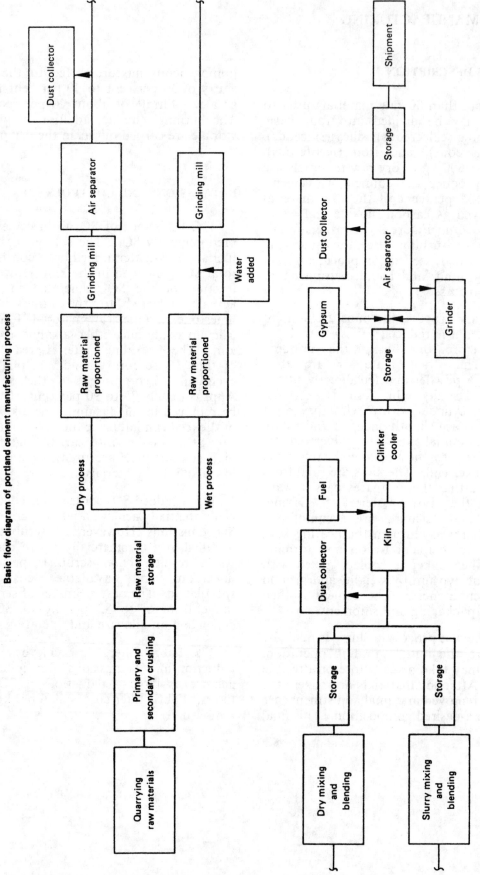

FIGURE 3

Basic flow diagram of portland cement manufacturing process

IV. METALLURGICAL INDUSTRY

A. IRON AND STEEL MANUFACTURING

377. The transformation of iron ore into steel and various forms of steel products takes place in a series of distinctly different processing steps. The four principal steps are:

(a) Preparation of raw materials
   - ore preparation
   - sintering
   - coke making

(b) Raw iron manufacturing

(c) Raw steel manufacturing

(d) Finishing of steel products.

378. As illustrated in Figure 4, numerous kinds of residuals are generated at the different steps of the production process. Process-related SO_2 emissions occur particularly in the sintering and coking processes.

B. PROCESS DESCRIPTION (10)(11)(12)

(a) Sintering

379. In the sintering process, the fine ore particles are agglomerated into a porous mass suitable for charging into a blast furnace. The purpose of the sintering is, apart from agglomeration of the ore concentrate, to burn some of the sulphur compounds contained in the ore concentrate. The sintering machine is charged with a mixture of iron ore concentrate, coke breeze, and limestone which is ignited by gas flames. In order to achieve the high temperature necessary for the agglomeration to take place, air is drawn through the mixture with the help of large fans.

380. The air pollutants generated during the sintering process are primarily particulates and sulphur dioxide. The sulphur dioxide is generated in the sintering furnace and originates from the sulphur in the iron ore concentrate and the coke breeze. The range of variation for the raw load of sulphur dioxide is 1.0 to 12.0 kg/tonne of sinter.

(b) Coking

381. The coke, which serves as fuel and reactive agent in the raw iron manufacturing process, is produced by heating metallurgical coal in the absence of air to a temperature at which the major part of the non-carbon components of the coal (i.e. volatile matter, water and sulphur) are driven off. Since around 25 per cent of the coal is carburated during the coking process, large amounts of gas ($\approx$ 5,000 Nm^3 per tonne of coke) are generated.

382. Coke oven gas is a sulphur-containing fuel which, when burned, can be an important source of sulphur dioxide to the atmosphere. The amounts and types of gaseous sulphur compounds found in coke oven gas depend upon factors such as the ash and sulphur content of the coal coked and coking-process conditions. Most of the sulphur in the gas is found as hydrogen sulphide (H_2S). Other sulphur compounds, termed "organic sulphur", are also found, principally carbon disulphide, but also thiophenes, mercaptans, and carbonyl sulphide.

383. In the most common method of coke manufacturing, the by-product coking process, the coke oven gas is subject to treatment in order to remove by-products such as tar, ammonia, sulphur, light oil, and phenol. The treated coke oven gas is either used as fuel elsewhere in the plant, sold as town-gas, or flared off.

384. The stack gas from coke oven gas combustion contains relatively modest amounts of SO_2. However, total SO_2 emissions from the coking process are associated with both underfiring of coke oven batteries and combustion of coke oven gas in other combustion units within iron and steel mills (Figure 5). Approximately one third of the sulphur contained in coal charged to coke-ovens is transferred to coke-oven gas. The range of variation for the raw loads of sulphur dioxide from a coke plant is 0.2 - 5.0 kg/tonne of coke.

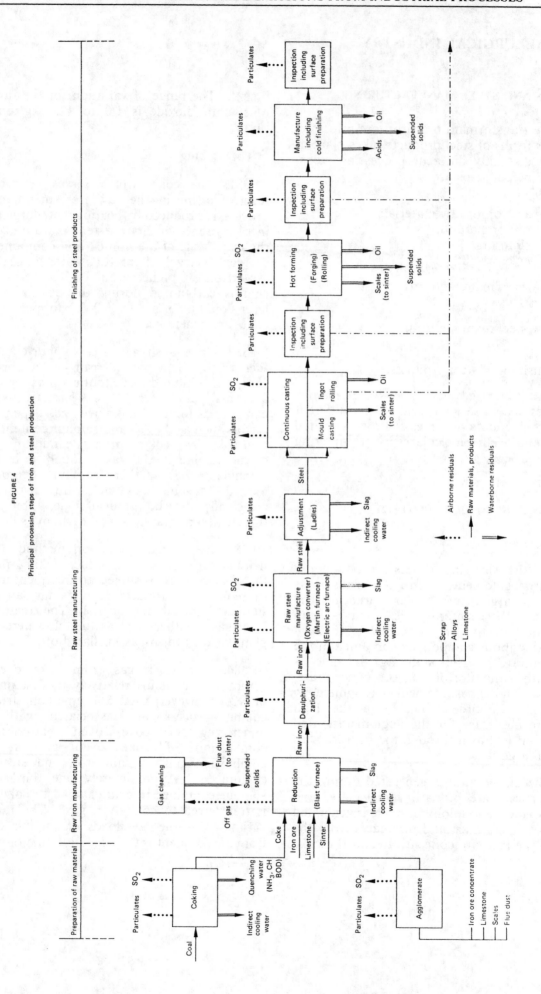

FIGURE 4

Principal processing steps of iron and steel production

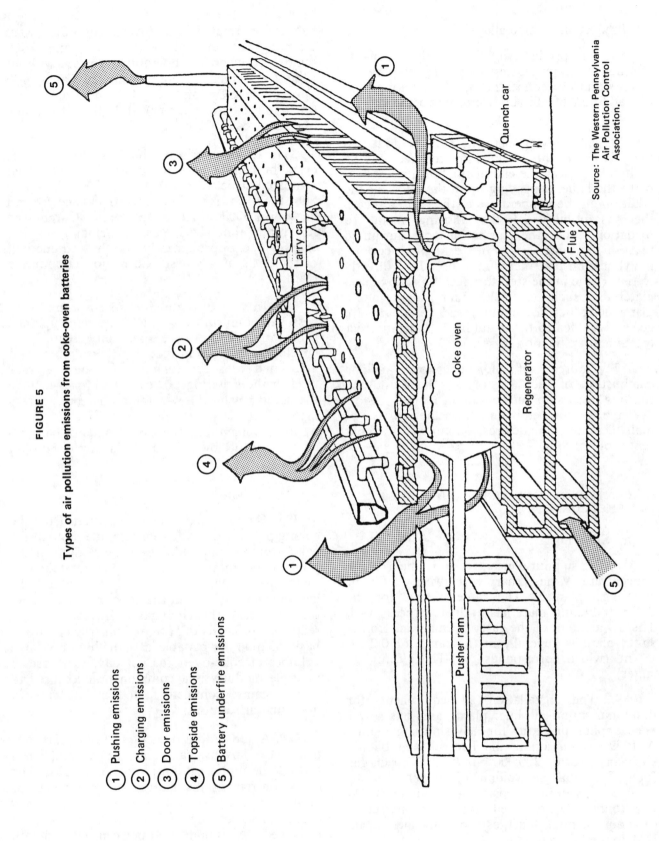

FIGURE 5

Types of air pollution emissions from coke-oven batteries

1. Pushing emissions
2. Charging emissions
3. Door emissions
4. Topside emissions
5. Battery underfire emissions

Source : The Western Pennsylvania Air Pollution Control Association.

(c) Production of ferro-alloys

385. The production of ferro-alloys consists of three basic processing steps: (i) preparation of ores and ore concentrates; (ii) smelting, and (iii) addition of alloying elements to the molten metal.

386. The alloying elements may be added to the charge in the steelmaking furnace, to the molten bath near the end of the finishing period, or to the ladle, depending upon the effect of the addition on the temperature of the molten metal, the susceptibility of the alloying metal to oxidation, and the formation of unwanted reaction products. For example, copper, molybdenium and nickel may be added directly to the charge in the steel furnace. Easily oxidized materials such as aluminium, chromium, manganese, boron, titanium and vanadium are generally added to the metal in the ladel in order to minimize oxidation losses.

387. Sources of SO_2 emissions include combustion of fuels for process heat and the roasting of sulphide-containing ores. For example, gaseous emissions from the roasting of molybdenium-sulphide ores contain 0.5 to 2.5 per cent (vol.) of SO_2.

C. CONTROL TECHNOLOGIES AND COST OF EMISSION CONTROL

(a) Sintering

388. The sulphur concentration in the iron ore concentrate varies sharply between different sintering plants. The same is true for the different fluxing materials and the process fuel. The SO_2 content of the gaseous emissions from a sinter plant is generally in the range of 0.2-3.0 g/Nm^3 (which corresponds to 1-12 kg/tonne of sinter).

389. The installation of equipment for desulphurization of sinter plant gases has so far been applied primarily for new sintering plants. A 1979 installation in the Federal Republic of Germany treating 320,000 Nm^3/h of SO_2-rich gas (out of a total gas volume of 800,000 Nm^3/h) incurred investment costs of around 15,500 DM/tonne SO_2 removed and operating costs (without capital charges) of around 2,000 DM/tonne SO_2 removed.

390. Under retrofit conditions where the whole gas volume rather than only the SO_2-rich gases has to be treated, the investment costs were estimated at around 30,000 DM/tonne SO_2 removed, whereas operating costs (without capital charges) were estimated at 6,700 DM/tonne SO_2 and total annualized costs (including capital charges) at 12,000 DM/tonne SO_2 removed (1979 prices).

391. In comparison, the costs of flue gas desulphurization at new power plants has been estimated at between 5,000 and 8,800 DM/tonne SO_2 removed regarding investment costs and between 1,500 and 2,600 DM/tonne SO_2 removed for total annualized costs (including capital charges), i.e. approximately one third to one half of the costs of gas desulphurization at sintering plants.

392. In view of the high costs associated with gas desulphurization at sintering plants, substantial amounts are spent on research and development aimed at reducing the sulphur input associated with the fuel and fluxing materials used in the sintering process. Examples include the development of low-sulphur coke breeze from lignite, reduction of the proportion of coke breeze by improvements in the overall energy balance, or the use of low-sulphur fuels in the ignition hoods.

(b) Coking

393. The principal gaseous emissions from the coking process are hydrogen sulphide (H_2S) in the coke-oven gas and sulphur dioxide (SO_2) in the products of combustion. The approach taken by most plants has been to try to reduce the concentration of H_2S in the coke-oven gas rather than to treat the flue gas. Installations for reducing emissions of H_2S are constructed on the basis of modules capable of handling very large volumes of coke-oven gas per day. The use of multiple modules is a common practice used to guard against complete failure in the case of operating disruptions.

394. A substantial reduction of SO_2 emissions from coke ovens can only be accomplished through desulphurization of coke-oven gas. The costs for retrofit installation of equipment for desulphurization of coke-oven gas have been estimated at around $US 4 per tonne of steel produced. Assuming a 90 per cent efficiency in overall sulphur removal, the average cost per tonne of SO_x removed corresponds to around $US 4,400.

(c)   Production of ferro-alloys

395. The emissions of SO_2 from ferro-alloy production depends on the sulphur content of the ore and ore concentrate as well as on the production technology. In order to control SO_2 emissions, a number of process modifications and add-on technologies have been applied. An example relating to the roasting of molybdenium-sulphide ore will be given below.

396. The gaseous emissions from the roasting of molybdenium-sulphide ores are first subjected to dust removal and thereafter fed into a sulphuric acid plant using a modified lead chamber process. The treatment consists of a multistage scrubbing of the SO_2/SO_3 containing gases with the addition of nitric acid (HNO_3) and nitrogen oxides (NO_x). This scrubbing results in sulphuric acid at a concentration of 70 per cent while sulphur removal efficiency is around 99 per cent. The gaseous emissions after treatment contain < 100 mg SO_2 per Nm^3 and 300 mg NO_x per Nm^3. Cost estimates made in the Federal Republic of Germany indicate that the process is economically feasible if the SO_2 content of the waste gas from the roaster exceeds 2.5 per cent by volume.

V. PULP AND PAPER INDUSTRY

A. GENERAL DESCRIPTION OF PRODUCTION PROCESSES (13)(14)(15)

397. The objective of all pulping processes is to separate the fibres in the wood from one another. This can be accomplished either by mechanical or chemical processes.

398. In the mechanical processes, the fibres are separated through grinding of the logs on grindstones in the presence of water. The result of this operation is a slurry consisting of more or less fully separated fibres. By the dewatering of this slurry on a close-meshed net, the fibres are matted together into a pulp sheet. Paper made out of such pulp is of limited strength and durability.

399. In the chemical processes, wood chips are treated with different chemical solutions in so-called digesters. During this operation, most of the wood's binding agent (i.e. the lignine) is extracted. In the sulphate process, lignine extraction is accomplished through the use of an alkaline digesting (cooking) liquor (primarily consisting of caustic soda). The sulphite processes use acid digesting (cooking) liquors (either calcium bisulphite or sulphites of magnesium, ammonium or sodium). The sulphate and sulphite processes based on magnesium, ammonium or sodium are all characterized by a high degree of recovery and re-use of chemicals whereas in the calcium-based sulphite process, no re-use of chemicals takes place.

400. In addition to these two basic types of production processes, there are a number of processing methods that combine elements of both types of processes. These so-called semi-chemical or chemi-mechanical processes basically contain the same operations as the chemical processes, the difference being that the wood chips are only partially digested and that the pulp therefore needs additional defibration in disc refiners.

401. The transformation of wood raw material into pulp or paper takes place in a number of distinctly different processing steps. The five principal steps are: (a) Preparation of raw materials; (b) Pulping (digesting); (c) Bleaching; (d) Pulp handling and/or paper production; and (e) Recovery of chemicals. These major processing steps consist, in turn, of a number of different unit operations which are shown in Figure 6. Whether all the different unit operations are actually carried out at a given plant depends on the technology of the plant and on the type of output.

Pulping

402. The purpose of the digesting operation is to let the digesting chemicals react with the lignine in the wood thereby accomplishing the desired separation of fibres. Except for some small amounts of spill, the major part of the chemicals used in the digesters are recovered and returned to the process. In the case of production of calcium-based sulphite pulp, there is no technically feasible way of reusing the chemicals. The spent liquor is therefore either burned (causing large SO_2 emissions), spray-dried to lignosulphonic powder, or discharged into the water.

Bleaching

403. The fact that many paper qualities require pulp of a high degree of brightness necessitates that the pulp be subjected to bleaching either with various kinds of chemicals (usually chlorine, sodium hypochlorite, chlorine dioxide, and caustic soda) in a number of steps, or with oxygen under high pressure and at a high temperature.

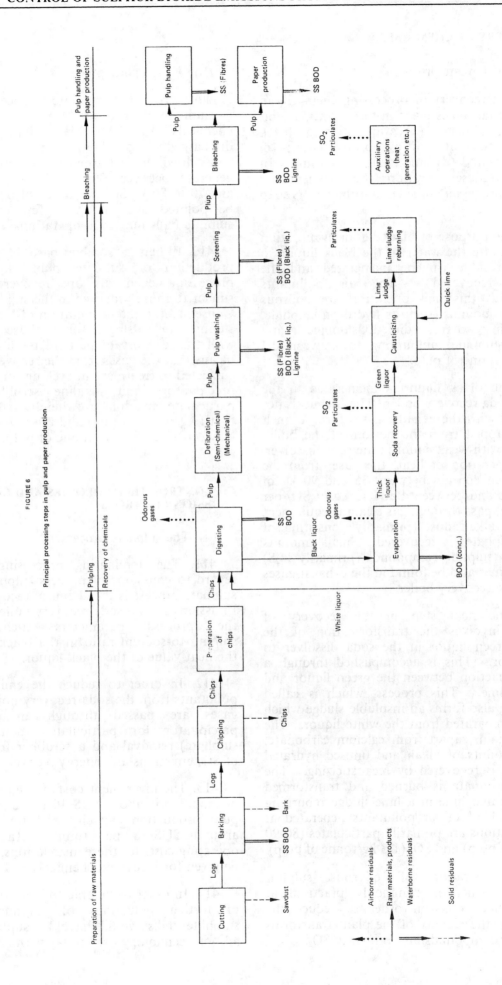

FIGURE 6

Principal processing steps in pulp and paper production

B. RECOVERY OF CHEMICALS

(a) The sulphate process

404. The recovery of processing chemicals in a sulphate mill takes place in four principal unit operations, namely: (a) evaporation of black liquor; (b) incineration of evaporated liquor; (c) causticizing, and (d) lime sludge reburning. In addition to these operations there is generally also some processing of such by-products as soap and resins.

405. The purpose of black liquor evaporation is to evaporate the water in the black liquor so that this residue can be incinerated and the chemicals recovered.   The air pollutants generated at this unit operation are odorous sulphur compounds (such as hydrogen sulphide, methyl mercaptan, dimethylsulphide and dimethyldisulphide) amounting to between 0.3 and 1.0 kg/tonne of pulp.

406. The black liquor is incinerated in the so-called soda recovery boiler. The chemicals not carried out with the exhaust gases form a smelt which is tapped from the bottom of the boiler and mixed with weak liquor in the soda dissolver. After recuperation of heat, the gases from the boiler, which contain between 35 and 90 kg of particulates and between 4 and 12 kg of SO_2 per ton of pulp, pass through gas cleaning equipment where the alkali dust (primarily consisting of sodium sulphate) is recovered. Small amounts of oderous sulphur compounds (primarily H_2S) can sometimes also be found in the exhaust gases of the soda recovery boiler.

407. The next step in the recovery of chemicals involves the transformation of the so-called green liquor in the soda dissolver to white liquor. This is accomplished through a chemical reaction between the green liquor and hydrated lime. This process, which is called causticizing also forms an insoluble sludge which has to be separated from the white liquor. The sludge contains, apart from calcium carbonate, certain amounts of alkali and unused hydrated lime which is recovered by recausticizing. The calcium carbonate is burned and transformed into dehydrated lime in a lime-sludge reburning kiln. The kind of air pollutants generated at these operations are primarily particulates (80-90 kg/tonne of pulp) and SO_2 (1-3 kg/tonne of pulp).

408. Incineration of odorous sulphur compounds which is commonly practised at many sulphate mills in order to reduce the nuisances in the vicinity of the plant transforms the sulphur compounds into primarily SO_2.

(b) The sulphite process

409. In the sulphite process by the calcium-based method, there is no re-use of chemicals and the black liquor is either discharged into the water or evaporated, spray-dried and burned.   Such incineration generates between 100 and 130 kg of particulates and 70 to 80 kg SO_2 per tonne of pulp. It should be pointed out that very few calcium-based sulphite mills remain in operation.

410. When a soluble base is used in the production of sulphite pulp, some of the processing chemicals are recovered in ways similar to those practised in the sulphate process. Consequently, the incineration of black liquor at sulphite mills using a soluble base is combined with the recovery of particulates and SO_2. This means that the gases from the recovery boiler are subjected to treatment in cyclones or electrostatic precipitators and alkaline scrubbers.   The particulate emissions are thereby reduced to less that 1 kg/tonne of pulp and the SO_2 emissions to between 10 and 30 kg/tonne of pulp.

C. CONTROL TECHNOLOGIES AND COST OF EMISSION CONTROLS

(a) The sulphate process

411. The dominating processing step with regard to emissions of process sulphur from the sulphate process is the chemical recovery system. Emissions from a soda recovery boiler depend on the process parameters such as the sulphur-to-sodium ratio in the black liquor and the heat value of the black liquor.

412. In order to reduce the emission of air pollutants from the soda recovery boiler, the flue gases are passed through an electrostatic precipitator for particulate (mainly sodium sulphate) removal and a scrubber for reduction of SO_2 emissions and energy recovery.

413. The investment costs for an electrostatic precipitator is about $US 10 to 12 per tonne of pulp production capacity and for a scrubber around $US 8 per tonne.   In calculating operating costs for these installations, credit must be given for recovered chemicals and heat.

414. In order to be able to meet requirements of further reductions of sulphur emission, sulphate mills would have to supplement the add-on technology used today with additional

measures, mostly consisting of process modifications. Examples of such measures are:

(i) Change of make-up chemical:

- Reduced input of bleachery waste acid;

- Use of non-sulphur make-up chemical;

- Use of low-sulphur fuel in the lime kiln.

(ii) Removal of sulphur from the chemical cycle:

- Removal of sulphur from the chemical cycle;

- Sulphonation of the pulp.

(b) The sulphite process

415. In the sulphite process, SO_2 and particulates are the main air pollutants. The control measures are, in principle, the same as for the sulphate process, i.e. evaporation and incineration of black liquor and recovery of chemicals, neutralization of spent liquor before evaporation, treatment of condensates, reduction of accidental losses, installation of precipitators and scrubbers for particulate and SO_2 removal.

416. With presently available technology, it is possible to limit total sulphur emissions from sulphite mills to around 2.5 kg per tonne of pulp. The corresponding figure for sulphate mills should be around 1.5 kg per tonne of pulp.

REFERENCES

1.   Airborne Sulphur Pollution: Effects and Control, Air Pollution Studies No. 1, ECE/EB.AIR/2 (United Nations publication New York, Sales No. E.84.II.E.8, 1984).

2.   Air Pollution Across Boundaries. Air Pollution Studies No. 2, ECE/EB.AIR/5 (United Nations publication New York, Sales No. E.85.OO.E.17, 1985).

3.   Guidelines for the Control of Emissions from the Non-ferrous Metallurgical Industries, Final Report of the Task Force. (United Nations Economic Commission for Europe, Geneva, 1979).

4.   "Emissions, costs and engineering assessment", Work Group 3B, Final Report, (June 1982, United States of America - Canada Memorandum of Intent on Transboundary Air Pollution).

5.   "Compilation of Air Pollutant Emission Factors, Third Edition (including Supplements 1-14)", (United States Environmental Protection Agency, AP-42, Office of Air Quality Planning and Standards, May 1983).

6.   "An Assessment of Reducing Emissions in Five Critical Industries for the Purpose of Acid Deposition Mitigation", (Energy and Resource Consultants, Inc., Boulder, Colorado, 1981) (Prepared for the Office of Technology Assessment, United States Congress, Washington, DC).

7.   "Assessment of Atmospheric Emissions from Petroleum Refining", Volume 5 (Appendix F: Refinery Technology Characterization, Radian Corporation, Austin, Texas, 1980) (Prepared for Industrial Environmental Research Laboratory, United States Environmental Protection Agency, Research Triangle Park, North Carolina).

8.   "Environmental management in oil refineries and terminals", United Nations Environment Programme, Industry and Environment, Vol. 8, (No. 2, April/May/June 1985).

9.   "Cost of control of sulphur dioxide, nitrogen oxides and particulates emissions from large combustion plants in oil refineries" CONCAWE, Report No. 7, 1984.

10.   Environmental Aspects of Iron and Steel Production - an Overview, United Nations Environment Programme, Industry and Environment Overview Series, 1984.

11.   The Making, Shaping and Treatment of Steel. W.T. Lankford and others, ed., United States Steel, (Herbick and Held, Pittsburgh, Pennsylvania, 1985).

12.   Emission Control Costs in the Iron and Steel Industry, Organization for Economic Co-operation and Develpment, (Paris, 1977). 13. Environmental Management in the Pulp and Paper Industry, Volume 1, United Nations Environment Programme, Industry and Environment Manual Series, No. 1, (Moscow, 1981).

14.   Pollution Abatement and Control Technology (PACT), Publication on the Pulp and Paper Industry, United Nations Environment Programme Expert Group Meeting on Information on Industry and Environment, (Paris, 14-16 January 1985).

15.   Environmental Guidelines for Pulp and Paper Industry, United Nations Environment Programme, Environmental Management Guidelines, No. 4, (Nairobi, 1984).

Chapter 6

EMISSION CONTROL BY FLUIDIZED BED COMBUSTION (FBC)

417. In recent years, the use of high-quality fuels for power and heat generation has dropped significantly owing to rapid increases in their prices. Correspondingly, the availability of alternative low-grade fuels, such as lignite, coal, and various kinds of industrial and municipal waste, has led to their increased use.

418. Combustion of these fuels in conventional boilers can, however, present serious environmental problems. These include flue-gas emissions of SO_x, NO_x, particulates and trace elements, some of which are highly toxic. Recently, considerable research and development work has been carried out worldwide in the field of fluidized-bed combustion as a means of overcoming the environmental problems associated with high sulphur fuels, in particular.

419. The Executive Body for the Convention on Long-range Transboundary Air Pollution decided at its first session in 1983 to carry out a special study on fluidized bed combustion, with the aim of evaluating the possibility of reducing sulphur and nitrogen oxides as well as other emissions in energy production while still satisfying users' demands for reliability and economy. With financial support from the Nordic Council of Ministers, the study was initiated in July 1984. A first draft was discussed at the fourteenth session of the ECE Working Party on Air Pollution Problems; this revision was then prepared on the basis of comments and subsequent information obtained.

I. STUDY METHOD

420. The study began on 1 July 1984; by 3 July 1984, some 46 questionnaires had been mailed to users, consultants, and manufacturers experienced with fluidized bed combustion (FBC). A further seven questionnaires were sent out later in July. The points covered in the questionnaires were:

    (a) Fluidized combustion device data

    (b) Tested emission performance

    (c) Availability performance

    (d) Cost data of the FBC plant.

421. Unfortunately, only two partly completed questionnaires were received. This is why some FBC manufacturers, users, and consultants were contacted individually for further information. Visits were made to plants in Finland, the Federal Republic of Germany, the United Kingdom, the United States of America and Sweden to obtain information on the following FBC units:

- Seven bubbling-bed FBC plants

- Four circulating-bed (CFBC) plants

- One multi-bed (MFBC) plant

- One pressurized fluidized bed (PFBC) plant.

422. All of the plants visited for this study had been in commercial operation for more than six months (except for the PFBC plant, which was a full-scale test plant). Small plants with thermal outputs of less than 10 MW were not considered relevant to the study. The results of the visits to the plants and discussions with the operators, manufacturers, and consulting engineers appear in this chapter. Unless stated otherwise, the data are from plants actually in operation and from users and manufacturers.

II. PROCESS DESCRIPTION

423. The burning of solid fuel is an old skill that has developed quite slowly with only a few major steps of improvement. Originally, the fuel was utilized in larger pieces, logs or coal chunks. Later, however, the invention of grate firing led to higher combustion efficiency using smaller particles. The next step was the introduction of pulverized fuel firing in the 1920s. This brought about flame combustion, already known from oil and gas firing, for solid fuels.

424. During the following 40 years, combustion technology was further refined, with the aim of raising efficiency and designing larger units. Today, for example, the biggest coal-fired boilers (for 1300 MW power stations) have a thermal output exceeding 3000 MW and a carbon burn-up of over 99.5 per cent. An important fact contributing to this excellent performance is the high combustion temperature. The modern furnace operates at flame temperatures over 1500°C, which is above the slagging temperature for coal. High availability on the other hand is dependent on the skill and alertness of the operators. Moreover, the boilers are highly specialized, which rarely allows different fuels to be burned simultaneously.

425. The most recent development in solid firing was the introduction of fluidized combustion. In the last 15 years, a significant amount of research and development has been carried out, leading to the successful application of this new technology for solid fuel firing. Although fluidized beds had been adapted to solid-gas reactions in chemiccal engineering for quite some time, their boiler applications have only recently become commercialized.

426. As opposed to flame combustion, high combustion efficiency is achieved in a fluidized bed at low temperatures, e.g., 800 to 900°C. This enables sulphur capture simply by adding an inexpensive absorbent such as limestone to the bed. The low temperature also significantly reduces the formation of nitrogen oxides.

427. Fluidization is generally understood to mean a non-stationary state in which solid particles are suspended by aerodynamic forces in a gas flow. Consider, for example, a bed of sand on a grate equipped with air nozzles. Increasing the pressure initiates a flow through the bed with the air first filtering through it. At a certain velocity, however, the bed expands and the particles start to move relative to each other in the flow. A distinct upper level is visible and an increase in the air flow does not correspondingly increase the drop in pressure. Rather the bed behaves like a liquid, e.g., when the vessel is tilted the level remains horizontal.

428. A further increase in velocity entrains the particles in the flow and then discharges them from the vessel. Thus another limit - entrainment velocity - has been reached. The region between these two limits is usally referred to as the bubbling bed. A continuous process cannot be maintained beyond the entrainment velocity unless the solids are separated from the offgas and circulated back to the bottom. This is precisely the principle on which the circulating fluidized bed works.

429. The bubbling bed is characterized by two separate zones: the bed proper with its high density of solids and the highly diluted freeboard above it. In the circulating case, the vessel is completely filled by the suspension, with a density similar to the bubbling bed at the bottom and gradually thinning towards the top.

430. When applying either of these principles to combustion, the aim is to ensure that the oxygen in the suspended air is consumed by the fuel. Both experience and calculations show that even for low heat values the ratio between fuel and inert bed material is very low, less than 5 per cent in fact. The fuel is suspended equally by the turbulent flow, which promotes a very efficient mixing of air and fuel. Larger fuel particles - from 5 to 30 mm - tend to stay longer in the bed, while fines may be carried away from the dense region. In both cases, however, the quality of the fuel particles is of minor importance for an efficient reaction. It is enough that they be combustible. Thus FBC is inherently a multi-fuel combustion technology. The FBC process is stable so that minimum supervision of the operation is required. The boilers are, in fact, well-suited for automatic operation.

431. Fuel preparation and feeding for FBC is in general uncomplicated. Compared to pulverized firing, no fuel grinding is required. Usually run-of-mine coal can be utilized as such, although if there are over 10 per cent of particles exceeding 35 mm in size, some crushing is recommended. There is no need for maintenance-intensive coal mills. In some designs, it is the pneumatic feed system and not the combustion that requires fuel to be crushed to less than 3 mm in size. In most of the plants visited, sulphur-absorbing limestone feed was mixed with the fuel into the combustion chamber.

432. The most recent development is to pressurize the FBC with the aim of reducing plant size and achieving a considerable increase in overall cycle efficiency using the combined steam-gas turbine cycle. The first full-scale pilot plant incorporating this idea, a 15 MW module that has been operating for some time, was visited in conjunction with this special study.

III. EXPECTATIONS AND CLAIMED ADVANTAGES OF FLUIDIZED BED COMBUSTION

433. Some of the advantages expected of fluidized bed combustion are listed below. A more detailed discussion of these points based on practical experience then follows.

Low Sulphur Emission Rate

(a) A low rate of S emissions is achieved simply by adding crushed or milled limestone/dolomite into the bed. Limestone decomposes thermally to CaO, which then reacts with sulphur dioxide and oxygen to form the end-product, $CaSO_4$ or gypsum. The most favourable temperature for this reaction is around 850°C. Owing to the thorough mixing and good contact between the bed material and the gases, even relatively moderate additions of absorbent will lead to effective retention of the sulphur.

Low Nitrogen Oxide Emission Rate

(b) Because the temperature of the bed is very low, almost no formation of thermal NO_x occurs. The fuel NO_x formation should be less than in conventional combustion because of the possibility of using staged combustion, especially in the circulating fluidized bed.

No Ash-fusion Problems

(c) Since the bed temperature is lower than the melting temperature of the ash, fusion problems should not arise.

Wide Fuel Range

(d) A wide range of fuels can be burnt in the same FBC reactor because of the inherent multifuel capability of the fluidized bed itself and the universal fuel-handling possibilities.

High Combustion Efficiency

(e) Fuels heat up rapidly when fed into the hot combustor mix. Combustion also takes place evenly and in a controlled manner. Any unburned particles are carried off by flue gas and can be easily recycled back into the furnace for complete burn-up.

High Coefficient of Heat Transfer

(f) Heat transfer between the solid wall and the suspension phase is known to be higher than between the solid wall and gas only. Thus it has proved possible to achieve very high heat fluxes through the heating surface immersed in the bed.

Space Saving

(g) A fluidized bed combustion boiler generally takes up less space than boilers with other firing systems yielding the same heat output. Space savings increase further when circulating fluidized bed combustion technology is used and dramatically so with pressurized fluidized bed combustion technology.

Good Operational Properties

(h) It is claimed that FBC is flexible in turndown and able to offer rapid start-ups and high load response rates.

High Availability

(i) Owing to the simple and rugged construction of the fuel and ash systems, along with the elimination of ash fusion, good availability of the system is expected.

Low Overall Investment Costs

(j) If flue gas desulphurization is included, an FBC plant always remains the cheapest option available.

IV. OPERATIONAL AND AVAILABILITY DATA OF THE PLANTS STUDIED

A. TYPE OF FUEL

434. An important advantage of FBC technology is its ability to burn different fuels, either alone or mixed. An excellent example of this multi-fuel capability is the CFBC plant that has burnt all of the following fuels with high combustion efficiency:

- Crushed coal

- Crushed brown coal

- Milled peat

- Bark

- Sawdust

- Paper waste

- Plastic waste

- Coal char

- Old tyres

- Battery shells

- Heavy oil.

435. The mechanical feed system enabled relatively large pieces of fuel - up to 60 to 70 mm in diameter - to be burnt. Usually plants are designed to use at least a minimum of two main fuels. This in turn enables plant owners to switch to the most economical fuel available at any given time. FBC plants are also capable of burning poorer fuels than other burning devices. These include ash-rich fuels (up to 70 per cent as content on a dry basis), wet fuels such as flotation tailings, industrial and sewage sludges, and fuels with a low energy/volume ratio (refuse, etc).

B. FUEL AND SORBENT SYSTEM

436. Instead of describing the details and functions of all the fuel/sorbent systems examined, this section will concentrate briefly on unit operations only.

Fuel Handling and Storage

437. Conventional fuel handling is used in most cases and associated problems have, of course, been experienced. Most of these stem from the fact that handling solid fuel is more difficult than oil or gas. Several plants have overcome their initial difficulties and are now running at an availability equal to or higher than that for pulverized fuel plants.

Fuel Preparation

438. According to users of FBC boilers, there are only minor problems with fuel preparation, and even these were similar for other boiler types. They included dimensioning and wear of the equipment and in certain cases arching of fuel in silos caused by wet fuel and freezing.

Fuel Feed

439. *Mechanical feed* - Mechanical feed using a conveyor followed by a rotary valve and gravity chute serves for overbed feeding in bubbling beds as well as for feeding into the non-mechanical seal in circulating beds. In both cases the number of feeding points is limited to one or two per bed. This system has proven its reliability. One particular advantage it offers is the extremely wide distribution of accepted fuel particles, even with dimensions up to several centimetres. Too few feeding points, however, in bubbling beds has also led to low combustion efficiency because of uneven fuel distribution.

440. *Pneumatic feed* - Pneumatic underbed feeding is used by some bubbling bed manufacturers. Before feeding, the fuel is milled, the size distribution normally varying from 0 to 3 mm. The number of actual feeding points varies from one to ten per bed (or from one to several per bed compartment). The main problems with pneumatic feed are plugging of the feed pipes owing to moist fuel and sintered fuel entering the bed. There were also slight but not too serious problems with erosion. Only one manufacturer has claimed no problems at all with pneumatic feed, although in this case only one feed pipe was used.

441. *Hydraulic feed* - One manufacturer has carried out tests with hydraulic feed, mixing milled coal with water. This gives a moisture content of 23 per cent. There is only one feed point and the feed system is underbed feeding. Preliminary tests have been encouraging.

C. COMBUSTORS

Bubbling Bed Combustors

442. There have been some minor problems with bubbling bed combustors. In one case the bed tended to sinter when burning coal and sewage sludge. The reason for this was incorrect measurement of the primary air required. The situation has been rectified.

443. Refractory erosion, caused by air leakage from the air duct into the bed, has also been a minor problem in one plant. In-bed tubes in some cases showed evidence of erosion. Operational experience, however, reveals that this problem can be eliminated with proper dimensioning and operating procedures.

444. There were no visible signs of corrosion in the plants - even on air-cooled distribution plates. Circulating Bed Combustors

445. Only two problems with erosion of furnace wall tubes were reported, both caused by faulty design and soon corrected definitively by simple redesign.

D. ASH HANDLING

Bed Ash

446. In some cases bed ash was screened and the coarse particles discharged from the system. In some plants combustion air was used to cool these particles. According to owners, the bed ash handling systems employed have operated without disturbances.

Hopper Ash

447. The high amount of unburnt carbon in the fly-ash from bubbling bed combustors requires hopper ash to be recycled back to the bed. In newer plants providing sufficient residence time, the need for recycling is smaller and in some cases non-existent.

Fly-ash

448. Fly-ash from electrostatic precipitators or baghouses is usually transported pneumatically to ash silos. Rotary valve erosion has sometimes been noticed. In general, FBC technology does not require any specially designed fly-ash handling equipment. The systems for pulverized-fuel fired boilers already commercially available have proven to be sufficiently reliable in FBC plants.

E. FOULING AND SOOT BLOWING

449. Because of the low temperature in the bed, neither ash nor sulphur sorbent melts in FBC combustors. Thus there are no slagging or fouling problems in well-operated FBC boilers. However, during burning of garbage or other uncontrolled mixtures, for example, unwanted low-melting constituents in the bed have sometimes caused agglomerations leading to plant shutdowns. Using design fuel, there are no such problems. The fouling of heat transfer surfaces in the combustor is non-existent because of the "sand blast" effect of the bed material. The gentle "nonstick" FBC fly-ash does not foul the superheater and economizer surfaces after the combustion chamber. The need for soot blowing in FBC plants is therefore always less than in conventional boilers. No plugging problems have been reported.

F. OPERATIONAL ASPECTS

Start-up

450. The purpose of the start-up procedure is to heat up the bed to the level required for safe ignition of the fuel. In bubbling beds three main start-up methods are used: underbed or overbed start-up burners or a special hot gas generator. During start-up, the bed height is in most cases lower than during normal operation. Many circulating units use overbed lances for initial solids heating and in-bed oil lances to reach the temperature level required for safe introduction of the solid fuel.

451. In some cases the bed is divided into several compartments so that the start-up can proceed compartment by compartment. An examination of some plants using this technique left the impression than further development was required. Compartmental division also aims at better turndown ratios which can be achieved using other means.

452. The normal start-up rate for FBC boilers is 100 °C per hour, which corresponds to an eight-hour cold start-up. In most cases it is not the combustor that is the limiting factor in the start-up procedure, rather, it is the allowable heating rate of the boiler's pressure parts; for instance, the drum. The refractories used in some plants visited allow heating rates up to 200°C per hour without risk of damage. This corresponds

to a cold start-up time of about 3.5 hours. After a short shutdown period, an FBC unit can be started immediately on coal again because of the high heat capacity and temperature of the slumped bed. An example of start-up procedure recording is shown in Figure 1.

Response Rate

453. The response rate or ability to change the load of the plant is very good and comparable with oil-fired plants. Typical measured response rates are 3 per cent to 10 per cent per minute upward and 7 to 25 per cent per minute downward. It should also be stated that, normally, the limiting factor is not the process or the combustor itself but the ancillary equipment, such as the fuel transportation system, regulating and control equipment, etc. An example of a response rate recording is given in Figure 2.

Turndown Ratio

454. Turndown ratios of FBC plants are good compared to other systems. By dividing the bed into several compartments, the turndown ratio can also be increased if necessary. Normal measured values are from 3:1 to 4:1, the best being 5:1. In some cases the high turndown ratio is achieved by increasing the amount of excess air. Reported turndown ratios are given in Figure 3.

G. AVAILABILITY AND PERSONNEL REQUIREMENTS

455. Not all the plants visited were willing to give operational availability data. It should be

V. COMBUSTION AND EMISSIONS

A. COMBUSTION EFFICIENCY

458. Manufacturers of FBC plants promise combustion efficiencies better than 95 per cent in most cases. During this study only relatively few measured data were received. Obviously many manufacturers of FBC combustors have had problems with combustion efficiency. The main factors affecting this are residence time, the degree of mixing, temperature, fuel quality, excess air, etc. In most cases, the FBC plant is equipped to recycle material elutriated from the combustor chamber and captured in hoppers and cyclone collectors located between the combustor chamber and the boiler heat-transfer surfaces. It is quite obvious that CFBC and PFBC plants are better than bubbling beds in respect to their

kept in mind that the technology in question was not yet fully developed at the time of design. The data obtained are presented in Figure 4. On average the performance is either the same or better than for conventional boiler plants.

456. As a result of the already mentioned suitability for automatic operation, manpower requirements are very low. One plant visited was part-time, manned during day shifts and run unmanned evenings and nights. A CFBC unit added to a paper mill boiler house required no additional crew: two men per shift operated all four boilers.

H. CONCLUSIONS

457. Because FBC technology is quite new, some difficulties in design and operation of the combustor and its ancillary equipment are to be expected. In most cases the problems have been easy to solve; only a few structural changes have been required. Most failures have been connected with fuel handling systems: plugging, freezing, erosion, and mechanical weaknesses. All these were, nevertheless, easy to correct. Ash-handling systems have also been improved to conventional standards. Based on the experience gained in FBC plants of various types and sizes, it can be stated that the availability of such plants is not lower than for conventional plants. As further experience is gained, operational availability is expected to increase.

combustion efficiency. The combustion efficiency of the CFBC plants investigated was higher than that of any other firing method except for pulverized coal firing. Some measurements results are given in Figure 5; in Figure 6, combustion efficiencies of pilot plant tests are presented.

B. SULPHUR RETENTION

459. Sulphur removal efficiency in FBC plants is, as a rule, very high when limestone or dolomite is used for sulphur absorption. The sulphur capture mechanism is complicated and the details not yet clear. Some computer programmes have been developed for theoretical

FIGURE 1

FBC plant start-up recording

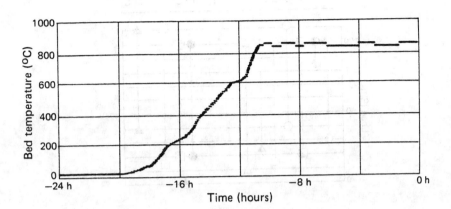

FIGURE 2

Response rate recording

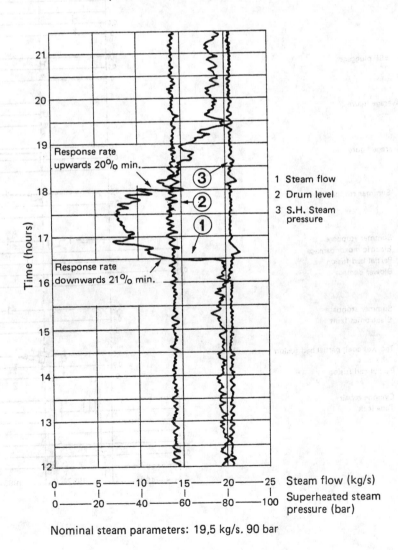

1 Steam flow
2 Drum level
3 S.H. Steam pressure

Nominal steam parameters: 19,5 kg/s. 90 bar

FIGURE 3

Turndown ratios of some FBC plants

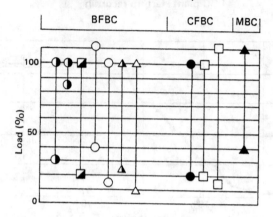

FIGURE 4
Availability data of some FBC plants

H Handed over to the customer
1 Change of all fabric filter hoses
2 Change of some fabric filter hoses
3 Revision

1 ESP pluggage

Average figure

Average figure

1 Summer stoppage

1 Summer stoppage
2 Electric motor damage
3 Partial bed fusion
4 Blower damage

1 Summer stoppage
2 Electronics fault

1 Too wet coal; partial bed fusion
2 Tube leak
3 Partial bed fusion

1 Cyclone repair
2 Tube leak

FIGURE 5

Measured combustion efficiencies of some FBC plants.

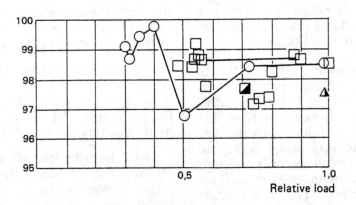

FIGURE 6

Measured combustion efficiencies in pilot plants.

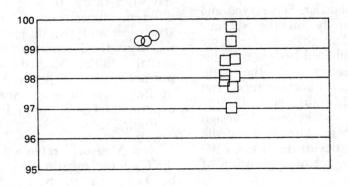

FIGURE 7

SO$_2$ retention guidelines for CFBC boilers from the same manufacturer

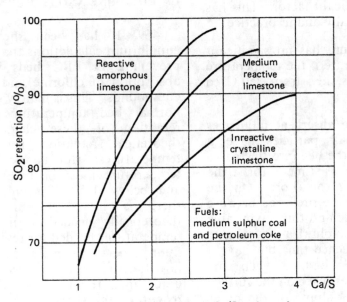

Sulphur retention versus Ca/S molar ratio

equilibrium calculations and some manufacturers also have programmes to calculate predicted retention rates. In many cases pilot plant or full-scale tests are made to predict the retention rate. The most important practical sulphur retention dependencies are listed below.

460. **Calcium to sulphur molar ratio** is obviously the most important parameter in sulphur capture reactions. The lower this ratio, the less absorption material needed and the less waste material produced. If dolomite - a mixture of calcium carbonate and magnesium carbonate - is used for sulphur retention, only calcium reacts with the sulphur-containing gases to form calcium sulphate or gypsum. Thus the amount of sorbent and consequently the amount of residue are considerably increased when using dolomite.

461. **Fluidized bed combustion boiler type and design level** impact strongly on the sulphur retention rate. It is evident that, at the same design level, circulating fluidized beds are in this respect better than bubbling beds the main reasons being the longer residence time and better distribution of the absorbent in the combustion chamber (see also below). ("Design level" in this case refers to the knowledge of the reactions and phenomena occurring in the FBC boilers as well as to the degree of application of this knowledge.)

462. **Air ratio** effects are neglible if the ratio is higher than 1.15. FBC systems will always operate at an air ratio of more than 1.1; therefore, the effect of the air ratio on the sulphur retention rate is an insignificant factor. This has been shown both theoretically and in practice.

463. **Reaction temperature** has no significant influence on the retention rate if the temperature is the same as the FBC's, i.e., from 750°C to 1000°C.

464. **Particle size of the absorbent** influences the retention rate: small particles have a tendency to elutriate from the bed into flue gases so that the residence time for the desulphurization reaction is too short. On the other hand, small particles are preferred because their specific surface is large and thus favourable for the surface reaction of sulphurization. Larger particles have enough residence time but have a big disadvantage in the bed: when the sulphurization reaction takes place on the surface of the particle, the molar volume of the outer layer of the particle increases in comparison to

the orginial particle material. This makes it very difficult for gases to penetrate, which in turn leads to low overall sulphurization efficiency of the particle material. Accordingly, it follows that for different beds different optimum size distributions exist for the absorbent.

465. **Residence time** of the reaction has a significant influence on the retention rate. This is confirmed by many theoretical, laboratory, and full scale tests. Residence time can be increased by recirulating the fly-ash and absorbent.

466. **Distribution of degree of mixing** of absorbent in the bed or the combustion chamber is an important factor in the retention rate. Good mixing, of course, improves the retention rate.

467. **Origin of limestone or dolomite** affects the retention rate. It is known, for example, that different limestones have different crystallic structures and also different degrees of porosity, which partly determines the retention rate. The actual influence played by the origin factor is normally tested either in pilot plants or directly in full scale plants. Some curves, based on experimental data from CFBC boilers, are shown in Figure 7.

468. **Measured retention rates** as a function of the Ca/S molar ratio are given in Figure 8. It can be stated that the best rates are those of CFBC and PFBC plants, those for bubbling beds being lower.

C. CHLORINE AND FLUORINE EMISSIONS

469. It has been shown theoretically by equilibrium calculations that at the temperatures prevailing in FBC beds there is almost no absorption of chlorine and fluorine by calcium compounds. This is mainly owing to the fact that normal bed temperatures are too high for retention of these elements. Considerable theoretical retention is achieved at lower temperatures: for 80 per cent retention, the temperature must be below 675°C for fluorine and below 425°C for chlorine, assuming that there is more calcium available than is theoretically needed. In practice, these low temperatures needed for retention of chlorine and fluorine occur when flue gases are cooled by the heat-transfer surfaces. As the cooling rate is quite high, the residence time for the capture reaction is too short; this is one reason why there is incomplete retention of chlorine and fluorine.

FIGURE 8

Measured SO₂ retention rates of FBC plants

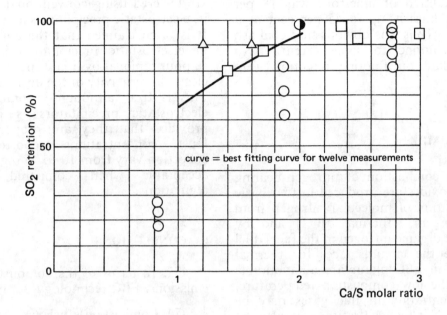

curve = best fitting curve for twelve measurements

FIGURE 9

Measured NOₓ emissions versus air ratio

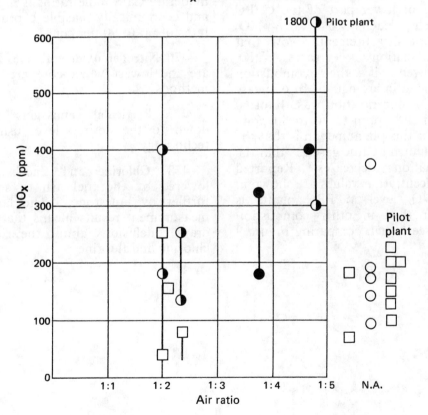

470. Data concerning measured chlorine retention were given by only one manufacturer. According to measurements, chlorine retention without any addition of limestone was 18 per cent, increasing linearly up to 45 per cent at a Ca/S ratio of 2. Thus it is obvious that coal ash also absorbs chlorine to some extent. No measurement data on fluorine retention were received.

D. NITROGEN OXIDES

471. During combustion of nitrogen-bearing fuels, nitrogen oxides are produced in two basic ways: by oxidation of molecular nitrogen from the combustion air ("thermal NO_x") and by oxidation of the organic nitrogen in the fuel ("fuel NO_x"). Because the circumstances for thermal NO_x formation in FBC plants are unfavourable as a result of the low combustion temperature, low NO_x concentration in flue gases may be expected. Fuel NO_x concentration can also be lowered by using staged combustion and flue gas recirculation. In the staged combustion method, the combustion air is divided into primary and secondary air so that there is a shortage of oxygen in the bed or lower part of the CFBC combustor, which leads to low NO_x concentrations. Test measurements show that flue gas recirculation decreases NO_x concentrations, even if the combustion temperature and air ratio are equal. In one case the decrease in NO_x concentration was about 60 per cent down from 180 ppm to 70 to 80 ppm. One explanation for this phenomenon is that the combustion air is diluted by flue gases so that the fuel NO_x formation decreases. Reported measured NO_x concentration values are shown in Figure 9. The NO_x level in FBC plants is considerably lower than in other combustion plants, as can be seen when comparing Figure 9 with Figure 10.

E. PARTICULATE EMISSIONS

472. Particulate emissions from FBC plants are lowered using conventional technology, that is, electrostatic precipitators and baghouse filters. It is often claimed that the collection efficiency of electrostatic precipitators is reduced when sulphur removal by limestone is used because the electrical resistivity of the dust particles increases. This is true, although in practice the use of electrostatic precipitators is fully acceptable provided that they are properly dimensioned. Dust concentrations reported after dust collection vary from 10 to 45 mg/m³, that is, fully acceptable when compared with existing regulations.

F. CONCLUSIONS

473. In terms of the combustion process and emissions, FBC technology is advanced.

474. **Combustion efficiency** when burning coal is high - higher than that of grate firing but somewhat lower than for pulverized firing.

475. **Sulphur retention** from flue gases with moderate Ca/S mole ratios is high, ensuring easy and economically feasible capture of sulphur at rates of up to 90 per cent.

476. **Nitrogen oxide emissions** from FBC plants are the lowest when compared to other firing methods.

477. **Particulate emissions** can be brought down to the desired level using conventional technology.

478. **Chlorine and fluorine emissions** are lowered by the fuel ash to some extent and further by limestone. The lack of sufficient measurement results means that it is difficult to draw conclusions about the retention rate of chlorine and fluorine.

FIGURE 10

Baseline NO$_X$ emissions of different firing systems

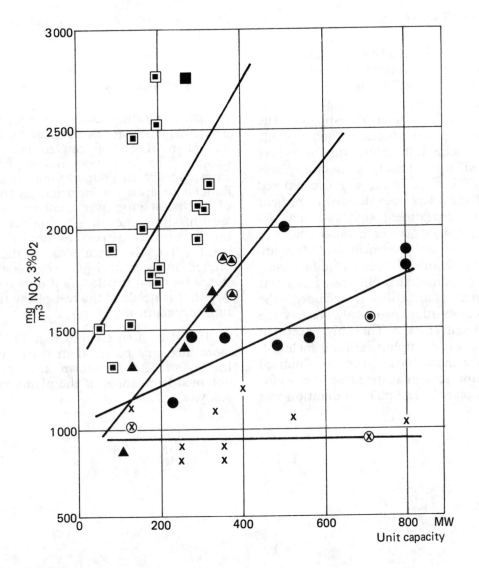

Federal Republic
    of Germany USA (pre NSPS^a)

     ⊕   ▲   front-fired

     ◉   ●   opposed-fired

     ⊗   X   tangential-fired

     ▣   ■   wet bottom

a NSPS = New Stationary Sources Performance Standards

VI. COST DATA

A. INVESTMENT COSTS

479. In many cases it was either difficult or impossible to get information about the investment costs of a plant. If the data were given, difficulties arose in finding ways of comparing them with those of other plants. The type of plant also posed problems: some plants produce steam, some hot water, while in others the price of auxiliary fuel peaking equipment was included, etc. The value of the site also differed by country. Thus it has been decided to present in Figure 11 the investment cost data actually received from plant owners or manufacturers as such. Although exact investment cost data are missing, by consensus all the manufacturers, plant owners and consultants but one agreed that "if flue gas desulphurization is included, the fluidized bed combustion plant always has the lowest investment costs". The only exception was that "if flue gas desulphurization is included, then the investment costs for a fluidized combustion plant remain at the same level as for a conventional plant". The plant in question was relatively small with a thermal output of about 20 MW.

B. OPERATING COSTS

480. Operating costs, consisting mainly of personnel costs, are as a rule at the same level or less than those of conventional combustion plants. The legislation and requirements vary from country to country, thus making it more difficult to draw a conclusion on the superiority of one plant type over another. Of course, the level of automation has a considerable influence on the number of persons needed to operate a plant. To give some idea of the possibilities offered by present FBC technology, it can be stated that in a smaller plant only one person per day shift is needed; the rest of the time the plant runs automatically.

481. It is often claimed that FBC plants need more auxiliary power than conventional plants. In Figure 12 are shown the measured data obtained from some of the plants visited for the study.

FIGURE 11

Investment costs of FBC plants versus thermal output (December 1984)

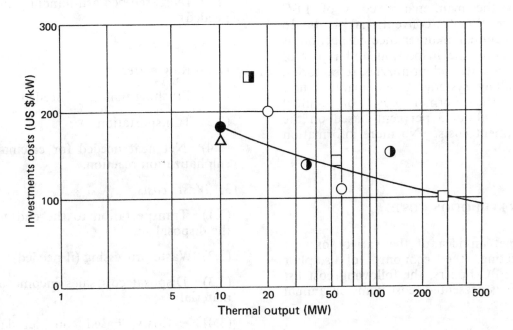

FIGURE 12

Auxiliary power demand of FBC plants versus thermal output

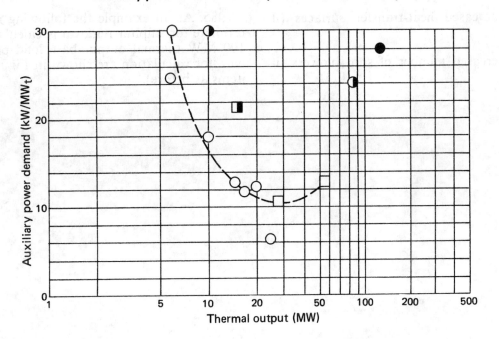

C. MAINTENANCE COSTS

482. Most users and manufacturers are of the opinion that the maintenance costs of FBC plants are lower than for conventional plants. In general only annual maintenance inspection is required, covering the inspection and repair of the distribution plate and air nozzles, refractories, and fuel-handling systems. One manufacturer states that the normal cost of annual maintenance is about 3 per cent based on the plant's investment costs. No more information was given.

D. SULPHUR REDUCTION COSTS

483. To get an idea of the variety of cost factors affecting the economy of sulphur reduction in FBC boilers, the following cost list of main items is presented (cost items on annual basis)

(1) *Discounted installed costs (discounted equipment costs plus installation costs)*

(1.1) Sorbent crusher (if needed)

(1.2) Sorbent storage bin

(1.3) Sorbent feed and control system

(1.4) Increased heat-transfer surfaces (if needed)

(1.5) Increased number of soot blowers (if needed)

(1.6) Increased fly-ash removal capacity (if needed)

(1.7) Increased bed ash-handling capacity (if needed)

(2) *Sorbent costs*

(2.1) Raw material

(2.2) Crushing/milling (if needed)

(2.3) Transportation

(2.4) Net heat needed for calcination and sulphatization reactions.

(3) *Waste costs*

(3.1) Transportation to the end-user or to the disposal site

(3.2) Waste processing (if needed)

(3.3) Disposal cost/sales income as a raw material

(3.4) Costs based on legislation or stipulations issued by authorities (if any)

(3.5) Other costs.

484. In view of the above list, it is easy to understand how impossible it is to arrive at universal cost amounts, especially when many of the plants visited gave out no cost figures.

485. As an example the following rough cost estimate for sulphur removal is calculated for an 100 MW thermal output base load plant (basic data for calculation are shown in Figure 13, cost items as before):

Figure 13

ROUGH COST ESTIMATE FOR 90 PER CENT SULPHUR REMOVAL OF 100 MW THERMAL OUTPUT BASE LOAD FBC PLANT (February 1986)

	Annual cost (US$ per annum)	Percentage
(1) Discounted installed costs:		
10% interest, 5 year pay-back time		
1.1 Sorbent crusher: not needed (see 2.1)		
1.2 Sorbent storage bin for five days' peak load use (100 m³) $ 15,500	4 095	1.9%
1.3 Sorbent feed and control system including SO_2-analyser $ 29,000	7 675	3.6%
1.4 - 1.7 not needed in this type of FBC boilers (based on experience)	-	-
(2) Sorbent costs:		
2.1 Raw material: limestone 0-1 mm in diameter, 20 $/t for 6,700 t	134 000	62.4%
2.2 Crushing: not needed	-	-
2.3 Transportation: 100 km by truck, 8 $/t for 6,700 t	53 600	24.9%
2.4 Net heat needed for calcination and sulphatization - at the Ca/S ratio of 2 this is negligible	-	-
(3) Waste costs:		
3.1 Transportation of the extra ash owing to sulphur removal: 20 km by truck, 2.5 $/t for 6,200 t	15 500	7.2%
3.2 - 3.5 not taken into account due to different circumstances in different countries	-	-
	214 870	100.0%

TOTAL COSTS: US$ 234 per tonne sulphur removed
US$ 0.286 per MWh net heat

NOTE:
In order to calculate sulphur removal costs per electric unit, the above-mentioned amounts must be divided by the overall plant electric efficiency (from coal to electricity in the network) and multiplied by the assumed FBC boiler thermal efficiency shown in Figure 13 (equal to 0.92).

EXAMPLE:
Overall electricity efficiency = 0.32
Total sulphur removal costs: US $ 234 per tonne sulphur removed
US $ 0.822 per MWh net electricity

486. It must be pointed out that the case described above is only one example (however realistic) which clearly shows that the most important cost items are those connected with the sorbent, i.e., the cost of sorbent raw material and the cost of sorbent transportation - which together represent more than 87 per cent of the total costs of the example case. Because the price of limestone varies greatly from country to country, the costs of sulphur removal also varies considerably. Compared to any other flue gas desulphurization method, the FBC boiler as a sulphur reducer seems to be the cheapest one. It must also be kept in mind that the waste from FBC boilers in some countries is not classified as a hazardous pollutant so that it can easily be disposed of on land or even in seas or lakes. In some other countries the end-product can only be disposed of in licensed waste disposal sites. It can also be used as raw material or construction material for industry.

E. CONCLUSIONS

487. The amount of cost data received from the manufacturers and users of FBC plants is both small and narrowly based. The main conclusion that can be drawn is that both the investment costs and overall costs are either on the same level or less than for conventional plants if flue gas desulphurization is included.

VII. FUTURE TRENDS AND PERSPECTIVES IN FLUIDIZED BED COMBUSTION

Bubbling Beds

488. The heat output of bubbling beds seems to be limited to the range 100 to 150 MW_{th} per bed owing to the relatively low heat-release rate of the bed area. This leads to large beds with their complicated and costly consequences. The reliability of bubbling beds at the technical level so far attained is good and their environmental features fully acceptable. In its capacity range, the bubbling bed is economically competitive with other methods, and there is good reason to believe that it has a promising future for the burning of solid fuels.

Multi-beds

489. Multi-bed boilers equipped with two separate beds are at present limited to around 50 MW heat output. Based on the experience of one 10 MW_{th} plant, this type is reliable, its desulphurization capability good, and the investment required moderate. On this basis, it can be regarded as a good alternative to conventional burning devices.

Circulating Fluidized Beds

490. Circulating fluidized beds are the largest and most often built CFBC boiler under construction at present, being about 290 MW_{th}. They can be designed for up to 400-600 MW thermal output. According to one manufacturer: "There are no longer any limits on plant capacity". In CFBC boilers, flue gas desulphurization is easy to carry out in a cost-effective manner and the desulphurization residue is also easy to handle and dispose of.

491. The data given by plant users show that CFBC boilers are reliable in operation, their availability often being higher than for conventional plants. Economically, CFBC plants are competitive with any other plant types if flue gas desulphurization is included. On the basis of the above, it is clear that CFBC technology is suitable for large plants and readily available at present.

Pressurized Fluidized Beds

492. Experience of a 15 MW_{th} test plant shows that this type of plant is highly favourable from an environmental point of view: its desulphurization capability is reported to be the best available and its NO_x concentration low. One manufacturer offers PFBC plants up to 793 MW_{th}, corresponding to 332 MW_e carried out in a combi-cycle. Because there are no PFBC plants yet operating commercially, it is difficult to predict the future of this technology.

VIII. GENERAL CONCLUSIONS

493. The following conclusions may be formulated concerning the present state of FBC technology:

(a)   Fluidized bed technology has been successfully developed into a commercially viable alternative to other combustion technologies. At the same time, development work is continuing on the different bed types for even more cost-effective and environmentally sound energy-production plants that can use low-grade fuels.

(b) The capacity of FBC plants has grown to such an extent in recent years that, according to one manufacturer, there are no longer any limits. In actual fact, there are great differences between the various types of FBC plants in this respect, not to mention a lack of experience with large plants.

(c)   In spite of the new technology and its brief operational life to date, the availability of the best FBC plants has been excellent. This should assure the high availability of new large plants, since the experience gained so far can be applied to their planning and design.

(d)   Emission control in FBC plants has proven to be both easier and less expensive than in conventional plants. Reduction of sulphur emissions in particular is both easy and cheap. In some cases it can even be put to good economic use. Nitrogen oxide emissions are also controlled at no extra cost as a result of the low combustion temperature. Moreover, fluoride and chlorine emissions are also reduced noticeably with FBC technology. Particulate emissions can be controlled by conventional methods.

(e) The economy of plants using fluidized bed technology is at least as good as that for the best of plants using other technologies if flue gas desulphurization is included, based on efficiencies as shown in Figures 8 and 9 and without special treatment costs. FBC technology is competitive with existing dry additive desulphurization methods for the performance efficiency reached.

FIGURE 13

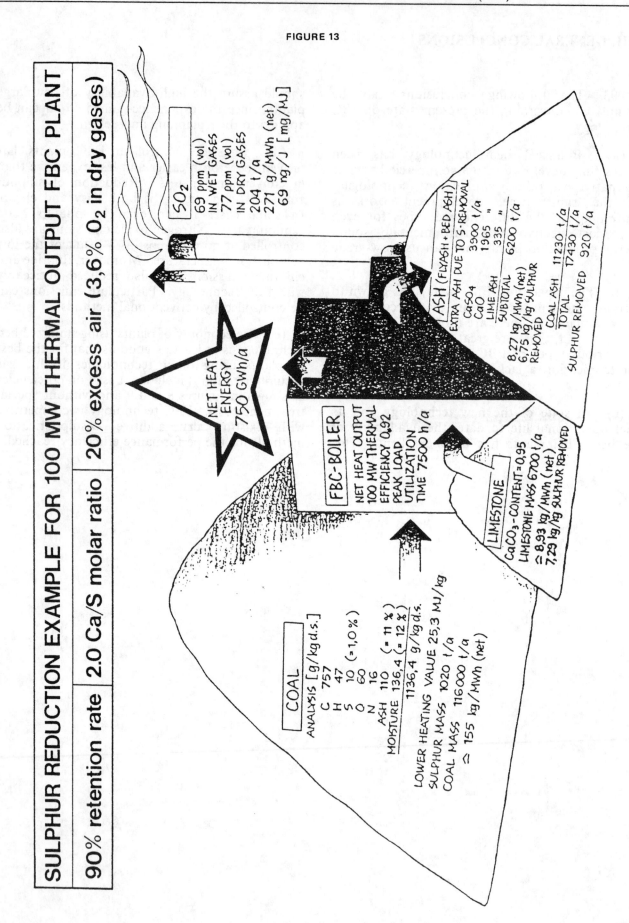

SULPHUR REDUCTION EXAMPLE FOR 100 MW THERMAL OUTPUT FBC PLANT

90% retention rate | 2.0 Ca/S molar ratio | 20% excess air (3,6% O_2 in dry gases)

SO₂
69 ppm (vol) IN WET GASES
77 ppm (vol) IN DRY GASES
204 t/a
271 g/MWh (net)
69 mg/J [mg/MJ]

NET HEAT ENERGY 750 GWh/a

FBC-BOILER
NET HEAT OUTPUT 100 MW THERMAL
EFFICIENCY 0,92
PEAK LOAD UTILIZATION TIME 7500 h/a

ASH (FLYASH + BED ASH)
EXTRA ASH DUE TO S-REMOVAL
CaSO₄ 3900 t/a
CaO 1965 "
LIME ASH 335 "
SUBTOTAL 6200 t/a
8,27 kg/MWh (net)
6,75 kg/kg SULPHUR REMOVED
COAL ASH 11230 t/a
TOTAL 17430 t/a
SULPHUR REMOVED 920 t/a

LIMESTONE
CaCO₃-CONTENT ≈ 0,95
LIMESTONE MASS 6700 t/a
≈ 8,93 kg/MWh (net)
7,29 kg/kg SULPHUR REMOVED

COAL
ANALYSIS [g/kg d.s.]
C 757
H 47
S 10 (≈ 1,0 %)
O 60
N 16
ASH 110 (≈ 11 %)
MOISTURE 136,4 g/kg d.s. (≈ 12 %)
1136,4 g/kg d.s.
LOWER HEATING VALUE 25,3 MJ/kg
SULPHUR MASS 1020 t/a
COAL MASS 116000 t/a
≈ 155 kg/MWh (net)

Chapter 7

INVENTORY OF TECHNOLOGIES FOR CONTROLLING EMISSIONS OF NITROGEN OXIDES FROM STATIONARY SOURCES *

494. National emission data of the past few years show that total emissions of NO_x from stationary sources parallel fossil fuel consumption. The proportion of fossil fuels used in primary energy production varies widely among countries because of the use of alternatives such as hydro-electric and nuclear power. The main stationary NO_x sources are boiler installations for the generation of electricity and other industrial uses. Depending on the structure of industry in individual countries, other relevant stationary NO_x emission sources may comprise specialized fuel combustion processes such as distillation furnaces, iron/steel plants, cement kilns, glass smelters, as well as certain non-combustion processes (e.g. nitric acid production).

495. In view of the fact that NO_x emissions from stationary sources arise primarily during combustion of fuel for generation of heat and electricity, the main emphasis is on a description of NO_x reduction technologies in large-scale power plants (i.e. power plants with a rated capacity of more than 50 MW^{th}) based on information available at the end of 1985. In countries with a quite different energy-supply structure or a large share of emissions originating from industrial processes, control strategies might differ and therefore necessitate other data.

496. NO_x is produced during the oxidation of fuels within a furnace at a rate governed by the characteristics of the fuel and the combustion conditions. Reductions in emissions may be made by suitable manipulation of the stoichiometry/temperature profiles within the boiler (primary measures) or by treating the flue gas at a later stage in its progress through the boiler (secondary measures).

497. Coal usually contains 1 to 2 per cent fixed nitrogen while commercially available residual oil contains up to 0.3 per cent N by weight. Moreover, combustible by-products of certain industrial processes may be contaminated with nitrogen compounds (e.g. waste gases in the chemical industry). Fuels such as natural gas and distillate oil are practically free of fixed nitrogen. Certain chemical reactions lead to the formation of nitrogen oxides in non-combustion processes such as nitric acid production.

498. The use of certain fuels is often determined by the energy supply structure of a country. Thus, the free choice of low nitrogen-containing fuels may be limited. Fuel cleaning with the sole aim of nitrogen removal is not always economical because of the technically complex requirements. Therefore, only a small reduction in NO_x is achieved by fuel treatment.

499. For new installations, several characteristics for low-NO_x emissions can be introduced satisfactorily at the design stage. However, the combustion systems of existing plants have usually been designed and operated on the basis of both technical and economic optimization of the process as a whole. In many ECE countries, boilers were built without regard for NO_x emissions; therefore, in the case of modifications made now for pollution control, possible adverse effects on performance have to be dealt with.

---

* Summary and conclusions of the report of a task force under the auspices of the Executive Body for the Convention on Long-range Transboundary Air Pollution. The full report of the task force was published by the Institute for Industrial Production, University of Karlsruhe (Federal Republic of Germany), in July 1986.

500. The intrinsic NO_x emission characteristics of a boiler are dependent upon several factors. Some are introduced at the design stage (e.g. wet/dry bottom, wall/corner firing, burner-zone heat-release specification, oil/coal/gas firing); some elements of the design may be modified fairly easily after construction (e.g. the use of different burner designs); other factors cause fluctuations in instantaneous NO_x concentration owing to variations in operating parameters (e.g.

load, excess air, coal composition, state of boiler-fouling, burner tilt (for tangential units)). Reduction of NO_x by combustion modifications usually involves the latter two categories.

501. Categorization of uncontrolled emissions from existing power stations is complicated by their variation with size, boiler design, fuel and operating conditions. Uncontrolled values usually fall into the following broad classes:

| Fuel/Firing | NO_x Emissions (mg/m^3) | | | |
	Gas (3% O_2)	Oil (3% O_2)	Hard Coal a/ (6% O_2)	Lignite/Brown Coal (6% O_2)
Tangential			670 - 1 000	400 - 800
Opposed	250 - 1 400	500 - 1 400	830 - 1 500	400 - 800
Single-wall			830 - 1 670	
Wet-bottom	-	-	1 000 - 2 330	-

a/ Emissions from larger coal-fired boilers usually fall into the upper range.

502. The site-specificity of emissions also implies that the extent of NO_x reduction possible by the retrofitting of any given technique is likely to depend on the design and operating parameters of the unit concerned.

503. Depending on site-specific parameters, NO_x reductions of up to 20 per cent can sometimes be achieved by minor modifications of the combustion process such as operating at lower excess air or by adjusting the fuel/air ratio of selected burners. However, the main area of interest for combustion modifications for NO_x control lies in the use of:

- Low-NO_x burner (LNB);

- Off-stoichiometric combustion (overfire air) (OSC);

- Flue gas recirculation (FGR),

all of which can be used either separately or in some combination. Major changes are sometimes required in order to implement such technologies as retrofits, although all are applicable to new units.

504. Low-NO_x burners (LNB) are well-developed, especially for boiler installations. Many well-established burner systems are

available for coal-, oil- and gas-fuelled boilers. For new facilities, the NO_x reduction attributable to LNB is about 30 to 60 per cent. Site-specific conditions can lead to lower reduction rates on retrofit applications. Installations of LNB's started in Japan and the United States of America in the 1970s. Beginning in the early 1980s, LNB's have been increasingly installed in new and existing European boilers.

505. LNB's have also been developed for other combustion processes, e.g. refinery furnaces or for the iron and steel industry, but their state of development and application is not so far advanced as for boilers. Moreover, the extent of NO_x reduction can be at variance because of process characteristics.

506. Off-stoichiometric combustion (OSC) usually refers to oxygen-deficient combustion of the fuel in the lower part of the firebox with a downstream addition of air. This technique is applicable to new and retrofit systems of all boiler types, depending on design. NO_x reduction ranges from about 10 to 40 per cent, depending on fuel and boiler type. OSC has been operated at a number of new retrofit large boiler installations in Japan and the United States of America since the 1970s. Practical experience in Europe has come about only during the past few years. Possible negative side-effects include boiler corrosion by reducing atmospheres

which might limit retrofit possibilities. OSC has also been tested at refinery furnaces.

507. Flue Gas Recirculation (FGR) is a technology particularly appropriate for gas- and oil-fired plants as well as for high-temperature coal combustion (wet bottom firing). About 10 to 20 per cent of the partially cooled flue gas is recirculated into the combustor. NO_x reduction can be achieved of about 20 per cent for coal, 20 to 40 per cent for oil, and up to 50 per cent for gas. This technology is well proven for new and retrofit installations although in the latter case its applicability may be limited by flame instability and problems with heat absorption in the furnace.

508. A combination of the above-mentioned primary measures means that a higher degree of NO_x reduction can be achieved than is possible by any single measure. The total reduction effect, however, is lower than the sum of the individual effects. Especially with retrofit systems the extent to which these methods can be used may be limited by possible side-effects such as slagging, fouling, enhanced corrosion, flame instability, efficiency losses and increases in other emissions. These side-effects have, however, been minimized in many cases by appropriate design and operation.

509. For existing utility boilers the following emission values have been demonstrated for retrofitting low-NO_x combustion systems:

(a) Pulverized coal firing (6% O_2)

wet bottom boiler:	1,000 - 1,400 mg/m^3	
dry bottom boiler:	600 - 800 mg/m^3	(tangential)
	600 - 1,100 mg/m^3	(wall-fired)
(b) Oil firing (3% O_2):	200 - 400 mg/m^3	
(c) Gas firing (3% O_2):	100 - 300 mg/m^3	

At new facilities emissions may often be lower than the smaller value in the above-mentioned emissions range.

510. Recent developments of primary NO_x control measures have focussed on further improvements of existing first- and second-generation low-NO_x burners, especially with regard to their flexibility in retrofit applications for uncommon fuels and boilers. Higher degrees of NO_x reduction (more than 60 per cent for coal firing) are expected from the operation of a new generation of burner with internal fuel staging. The same is true for the on-going development of a catalytic ceramic oil/gas burner. This technology is currently being implemented in Japan, the United States and the Federal Republic of Germany. However, full-scale operating data are still lacking.

511. Investment costs for primary measures are fairly low compared to those for secondary flue-gas treatment systems. These costs may be negligable for new plants but can range from 10 to 30 DM/kW$_{el}$ for a retrofit. Additional operating costs are claimed to be low in the majority of cases; although the data published have been insufficient to quantify them fully under all conditions.

512. Reburning is an emerging NO_x control technology which offers the potential of substantial additional reductions in NO_x. When used in conjunction with other combustion modifications, NO_x emissions may be thus lowered by up to 80 per cent. This technique consists of a second combustion zone in the boiler. Although some positive results have been reported, further development is required especially for coal-burning units. This system is probably not completely applicable to retrofits and cannot yet be regarded as proven technology in ECE countries.

513. When retrofitting low-NO_x combustion systems, burn-out problems increase with decreasing boiler capacity (below 100 MW heat input) for coal and residual-oil plants. Further investigations are being undertaken in order to improve the retrofit applicability of such technologies. At new facilities these problems can be reduced by appropriate boiler design.

514. Coal-fuelled grate and fluidized bed combustors usually exhibit lower NO_x emissions than pulverized coal boilers. For existing grate-fired plants NO_x emissions may be lowered to 300 or 600 mg/m^3 by excess air reduction, OSC and FGR, although reports of results are scarce

so far. For new installations it seems possible to achieve emission levels of 200 or 300 mg/m³.

515. Atmospheric fluidized bed combustors (AFBC) and ignifluid-systems have been built up to capacities of 300 MW$_{th}$. Two of the three main construction types, the circulating AFBC, and ignifluid have shown very low emissions and high combustion efficiency. NO$_x$ emissions of 150 to 300 mg/m³ have been measured. An additional advantage of the circulating AFBC is the high desulphurization efficiency achieved with the addition of basic sorbents such as limestone to the fuel. As for the other construction type, the bubbling bed AFBC, smaller units (particularly those below 20 MW$_{th}$) may have some burnout problems. NO$_x$ emissions are in the range of 250 to 700 mg/m³. In addition, desulphurization efficiency may be less than that of the circulating AFBC, although developments aim to improve its performance. Pressurized FBCs, which may have some further construction and environmental advantages over atmospheric FBCs, are still under investigation although a few plants are operating. Test results indicate that NO$_x$ emissions are in the range of 150 to 300 mg/m³.

516. Both investment and operating costs for FBCs can be somewhat higher than for grate-fired plants of comparable small capacity without flue-gas desulphurization. However, the potential of FBCs to burn cheap, low-quality fuels and the simple SO$_2$ removal technology can outweigh the disadvantages. The economic potential of FBC technologies for very large-scale power generation beyond 600 MW$_{th}$ is not yet quantified.

517. Low-NO$_x$ combustion systems have been developed in only a few cases for facilities other than boilers, refinery furnaces or steel heating furnaces. One example is the precalcination process for cement kilns which has the potential to reduce NO$_x$ emissions below 300 mg/m³ (10% O$_2$) while saving some energy at the same time. This process is operative in Japan and is now under investigation in some ECE countries.

518. For the combustion of nitrogen-containing chemical wastes, special low-NO$_x$ multi-stage combustors have been developed. Further investigations are aimed at the development of special low-NO$_x$ combustion systems for other industrial processes.

519. For stationary gas turbines, emissions may be up to 600 mg/m³ (15% O$_2$) although new low-NO$_x$ combustion chambers have been developed in some ECE countries. These have the potential to reduce NO$_x$ emissions down to 300 mg/m³ or even below 150 mg/m³ in some cases when firing natural gas. These gas turbines are now commercially available although retrofit applications are only possible in special cases. In addition, NO$_x$ reductions up to 80 per cent can be obtained by water or steam injection into the combustion chamber. Possible negative side-effects of the water injection are efficiency losses and an increase in smoke, carbon monoxide and hydrocarbon emissions.

520. For stationary internal combustion engines several primary measures for low-NO$_x$ operation are known, in particular the lean-burn concept, and other special combustion methods, exhaust gas recirculation, or water injection. Practical experiences with these technologies are still limited. NO$_x$ emissions down to 1,000 mg/m³ (5% O$_2$) or even lower seem to be possible.

521. Process modifications for low emissions of NO$_x$ in the operation of non-combustion sources such as nitric acid plants are widely known and put into practice. High-pressure processes have many advantages over low-pressure ones but modification of existing facilities is only possible in a few cases.

522. For flue gas denitrification at power plants, some technologies are currently offered commercially in the ECE region. The most highly developed process is Selective Catalytic Reduction (SCR). This process has been developed primarily in Japan for achieving up to 80 per cent NO$_x$ reduction. For all types of fuels this process uses ammonia to reduce NO$_x$.

523. For pulverized coal-fired boilers, three groupings of the SCR unit within the total flue-gas ducts are possible:

(a) High dust system;

(b) Low dust system;

(c) Tail gas system.

The High Dust System places the catalyst between the economizer and the air preheater, the flue gas having its original dust content. For the Low Dust System, the catalyst is in the same place but a high temperature electrostatic precipitator is located upstream. With the Tail Gas System, the SCR unit is arranged at the cold end of the plant; this necessitates re-heating of the cleaned flue gas to temperatures above 300 C.

524. The application of flue-gas treatment technologies in the ECE region will require some country-specific adaptations since optimization will have to take into account different regulations, fuel types, combustion systems, and

operating conditions. This applies mainly to coal-fired installations, since they need more modification than oil- or gas-fired plants.

525. The technical performance of the various flue-gas treatment processes and groupings is now being tested in a number of pilot plants in some ECE countries. More than 50 pilot plants are operating in the Federal Republic of Germany. These investigations are mainly aimed at finding the optimum design parameters for large technical plants under European conditions. In addition, commercial plants have started operation: in the United States one oil-fired power station is operating and in the Federal Republic of Germany two coal-fired power stations had commenced operations by the end of 1985.

526. At the end of 1985, more than 20 flue gas treatment units for an electricity generation capacity of about 7,000 MW were operating, under construction, or under contract in Austria, the Federal Republic of Germany, and the Netherlands. Operational data will soon become available. It is planned that both in Austria and in the Federal Republic of Germany almost all coal-fired power stations will be equipped with SCR or other flue-gas denitrification systems by the early 1990s.

527. Until now, the application of flue-gas denitrification measures had focussed on boilers and nitric acid plants. Increasing use of flue-gas denitrification for larger stationary internal combustion engines has been reported from Japan, the United States and the Federal Republic of Germany. For these engines, special SCR units have been developed and put into operation with a guaranteed NO_x reduction rate of 80 per cent. More than 90 per cent NO_x reduction is possible by using the well-known three-way catalytic converter concept, which can only be applied on four-stroke spark-ignition engines. According to Japanese publications, some special applications of SCR-technology in the iron and steel industry (sintering machine), for glass smelters and in refineries are known. Moreover, some furnaces in Japan and the United States are equipped with NH_3-injection systems without catalysts, which results in a lowering of NO_x emissions by 20 to 60 per cent. Current investigations in some ECE countries

have the objective of introducing flue-gas treatment technologies into other industries.

528. Some cost estimates for additional investment and additional total costs of an SCR plant for a 80 per cent NO_x reduction at new utility boilers in the Federal Republic of Germany are given in Figures 1 to 3. The cost of an SCR plant is influenced in particular by requirements of catalyst mass (depending on NO_x inlet concentration), catalyst lifetime (based on Japanese experience this can be assumed to be three to five years for coal, and five to seven years for oil and gas), on ammonia consumption, and by the annual full-load hours. The specific additional costs of electricity generation using an SCR plant for a power station with 700 MW_{el} at 4,000 full-load hours of operation per year (middle load range) are estimated to be:

for dry bottom firing	to 0.005 - 0.009 DM/kWh
for wet bottom firing	to 0.006 - 1.002 DM/kWh
for oil firing	to 0.003 - 0.005 DM/kWh
for gas firing	to 0.003 - 0.005 DM/kWh

Currently the costs for the tail-gas system fall into the upper range of the amounts given above, although this situation may change if a low-temperature catalyst can be developed.

529. Possible negative side-effects of flue-gas treatment systems include fugitive release of ammonia and other secondary emissions, enhanced deposition in air-heaters and downstream equipment, the interaction of ammonia with desulphurization plants and precipitators and the contamination of fly ash or waste waters with ammonia. The impact of these side-effects on cost will only be quantified after a substantial operating period under commercial conditions. Nevertheless, some ECE countries have taken the position that these consequences are of minor importance.

530. It is recommended that the actual efficiency of low-NO_x installations in the ECE region should be examined after a period of about two years in order to confirm current performance projections.

FIGURE 1

**Additional Investment for an SCR unit at new coal-, oil-,
and gas-fired utility boilers in the Federal Republic of Germany**

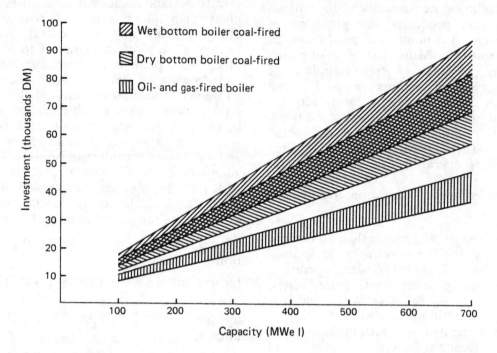

FIGURE 2

**Total additional costs of an SCR-unit at new coal-fired boiler
in the Federal Republic of Germany**

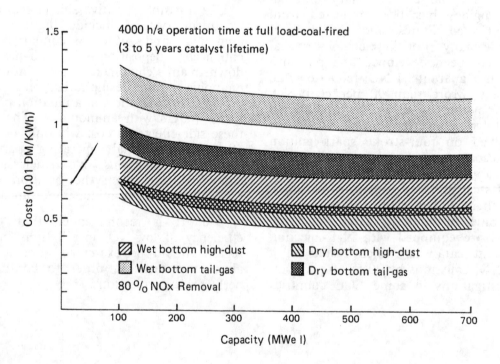

FIGURE 3

**Total additional costs for an SCR-unit at new oil- or gas-fired boilers
in the Federal Republic of Germany**

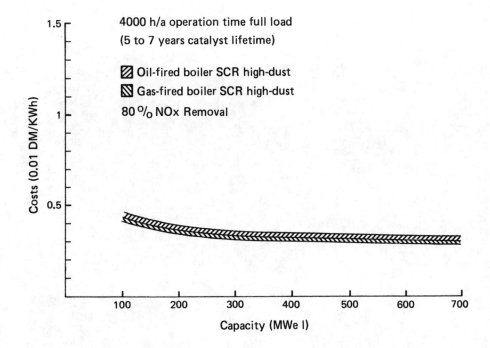